The focus of this series is general topics, and applications about, and for, engineers and scientists on a wide array of applications, methods and advances. Most titles cover subjects such as professional development, education, and study skills, as well as basic introductory undergraduate material and other topics appropriate for a broader and less technical audience.

# Synthesis Lectures on Engineering, Science, and Technology

Fariborz Barman

# Semiconductor Product Engineering, Quality and Operations

Deliver High Quality Products & Increase Profits

Fariborz Barman
Marvell Technology, Inc.
Santa Clara, CA, USA

ISSN 2690-0300 ISSN 2690-0327 (electronic)
Synthesis Lectures on Engineering, Science, and Technology
ISBN 978-3-031-18032-3 ISBN 978-3-031-18030-9 (eBook)
https://doi.org/10.1007/978-3-031-18030-9

This Springer imprint is published by the registered company Springer Nature Switzerland AG
The registered company address is: Gewerbestrasse 11, 6330 Cham, Switzerland

*I would like to dedicate this book to my mom and dad for providing me the opportunity to come to the United States and pursue my education and life dreams. I would also like to dedicate this book to my wife Barbara and my three wonderful kids, Julia, Myles and Victoria. Barbara, by doing almost everything else in our life, made it possible for me to focus on my career and become the best semiconductor engineer I could. My kids always encouraged me to try different paths and look at problems in different ways than my impulses would tell me to do. I also would like to recognize my dear siblings who encouraged me along my authorship journey.*

# Foreword

I spent 38 years in the design and development of complex data and telecommunication systems and over 24 years of it as an entrepreneur building two very successful bleeding edge semiconductor companies that I took public on NASDAQ and NYSE, Centillium Communications and Aquantia Corporation. Centillium was instrumental in setting the leading edge in broadband communication with its Digital Subscriber Line (DSL) semiconductor Integrated Circuits (ICs), while Aqunatia was the innovation leader developing high-speed Ethernet ICs for use in Data Center, Enterprise and Autonomous Driving automobile connectivity.

Semiconductors are the fundamental building blocks of everything electronics around us. It is stating the obvious that without the astonishing advancements in semiconductor design and manufacturing, there would be no Google, Amazon, Tesla, Meta, Instagram or any other of the digital constructs that define our current and future digital life. To give you a perspective on these advancements, Intel's 4004 processor introduced in 1971 had 2250 transistors in 10,000 nanometer process and occupied 12 mm$^2$. In contrast, as of 2021, the largest transistor count in a commercially available microprocessor is 57 billion MOSFETs, in Apple's ARM-based M1 Max system on a chip, which is fabricated using 5 nm technology and occupies 425 mm$^2$. That's over 134 million more transistors packed into a single IC which have to all work perfectly over time, temperature, manufacturing process variation and use cases, every single time. The often unsung heroes of this amazing feat are the engineers and scientist that have invented operations and manufacturing processes to deliver the highest level of quality, reliability and repeatability that allow the semiconductor industry to manufacture over 1.2 trillion ICs per year to keep the world's economy humming each and every day.

I have known Fariborz and the pleasure of working together at Aquantia Corp where he was responsible for test, quality and reliability of our products which were deployed by the leading data communications companies like Intel and Cisco with some of the most exacting standards. We worked on very high-speed semiconductor solutions which were being manufactured in the leading edge process technologies in different fabs with extremely short schedules. Fariborz's insight and discipline in semiconductor manufacturing processes is laid out in this book and is a must read for any young engineer planning to enter into this fascinating world of semiconductor operations. A world where your challenge is to find the one transistor out of billions fails at one corner of the process at one temperature corner under a specific mind boggling set of circumstances and then have the audacity to find cost-effective measure to devise processes and tests to prevent it from leaving your production line and on to the customer. The book is the result of practical knowledge collected over three decades of growth through many process technology innovations and learnings.

Fariborz takes you through the painstaking details of life cycle planning from before the chip design starts to bring up phase, characterization, qualification, production, cost reduction and maintenance. Each and every one of these steps in the life cycle is then broken down into its many important activities and analytics necessary to build confidence in shipping a high-volume product.

This is a must read and a necessary handbook for training future generations coming into this essential technology upon which our entire way of high technology life is standing.

San Jose, California, USA                                                   Dr. Faraj Aalaei

# Preface

Semiconductor Product Engineering and Operations presents a proven and efficient roadmap to deliver high quality, lowest cost possible semiconductor-based products in record time. The material will cover the entire semiconductor product life cycle from concept to obsolescence. The entire life cycle is segmented into several stages with PE responsibilities and best practices presented for each stage.

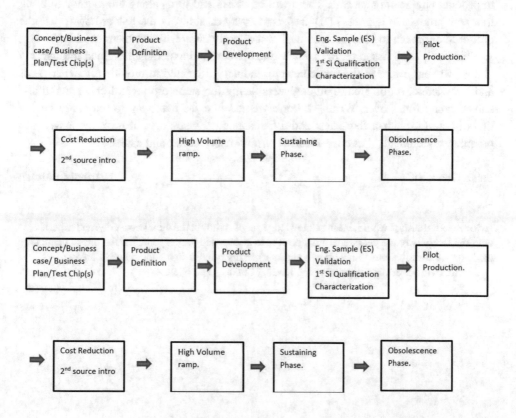

This work provides a central resource for semiconductor product engineers, quality engineers, operations professionals and management to refer to while navigating the complicated process of new semiconductor product introduction.

There are many books dedicated to different areas of the semiconductor industry such as digital and analog design, validation and test. I was never able to find a resource that addressed the need of semiconductor product, quality and operations engineers for a roadmap for how to successfully introduce new products.

A proven roadmap for a very successful new product introduction process for the chip industry. It covers best industry practices, JEDEC standards and other methods to put science in the forefront. The process has simplified process flow diagrams, success metrics, execution guidelines and checklists. The whole book is tailored based on a product engineering centric NPI hub and practical approaches.

Along with the core material, I have provided helpful hints, and best practices intended to help the reader to overcome common obstacles faced in critical stages of the new product introduction process.

Hypothetical product ZDD421A will be used to illustrate several points that are critical to product engineering success. There will be twists and turns along the journey to bring this new product to market. ZDD421A is my way of showing the reader how to use the proposed approach to identify "red flag" issues early and resolving them quickly.

This book is all based on my knowledge, opinion and experiences gained through my almost 40-year career in the semiconductor industry. I wanted to provide my experiences and knowledge in this area to all engineers facing the same problems I encountered and solved many times over. Writing a book seemed like the best way to reach my goal. I literally started with a few ideas and a blank page 4 years ago. It took me a while to organize my thoughts, but once I started writing it got easier and easier.

Santa Clara, USA                                                                              Fariborz Barman

**Acknowledgments**  I would like to acknowledge people who shared their knowledge and experiences with me freely over the years. I will not be able to acknowledge everyone, but the ones I name here made major contributions to my growth and sparked my interest in a wide range of areas.

David, Faraj, Sreenivas, Sunil, Parviz, Kwang, Sinan, Alex, Nick.

# Contents

# About the Author

**Fariborz Barman** has college degrees that include BS and MS in Computer Engineering both from University of California. I have been in the semiconductor industry since 1984 fulfilling several different roles with increasing complexity and responsibility as my skills and experience grew. I worked at companies such as National Semiconductor, Advanced Micro Devices, Quantum Effect Devices and a few more smaller companies. It has been an amazing journey, and I can't wait to share my knowledge and experience with you.

My first job was a process development engineer involved in the 1–2 um CMOS technologies. In the early days of my career, I was developing transistors and memory cells (i.e., the building blocks of every semiconductor product). Later on, I was responsible for transistor model parameter extractions which are the building blocks of SPICE transistor simulations tools. During the latter half of my career, I became interested in product and test engineering. So I used my knowledge of basics to move my career in that direction. More recently, I have managed new product introductions for several types of products including ASICs, FPGA and mixed signal products. I have direct experience with new products built using 10 um–28 nm process nodes.

# Abbreviations

| | |
|---|---|
| ASIC | Application Specific Integrated Circuit |
| ATPG | Automated Test Pattern Generation |
| BIST | Built-In Self-Test |
| CZTP | Characterization Test Program |
| DFT | Design For Test |
| ESD | Electrostatic Discharge |
| FPGA | Field Programmable Gate Array |
| HAST | Highly Accelerated Stress Test |
| HTOL | High Temperature Operating Life |
| MBIST | Memory BIST |
| RQTP | Reliability Qualification Test Program |
| SCAN | Modified flip-flops inserted in the design stage so it can be viewed as several segments or SCAN chains |
| STDF | Standard Test Data Format |
| Tape-Out | Submitting design database to wafer fabrication facility |
| TMCL | Temperature Cycling |

# List of Figures

# List of Tables

# Semiconductor Product Life Cycle, DFT and Budgeting

1

## 1.1 Life Cycle Chart and Hub of Influence Concept

A typical semiconductor based product has a life span of anywhere from 2 to 10 or more years depending on its features, popularity, customer adoption and market segment. This life cycle can be segmented into several distinctive stages as shown in Fig. 1.1

80% of a PE's efforts go into the first half of the product life cycle. During this stage the product goes from a concept through pilot production (or so called pre-production). This is a critical stage because of intense cycles of issue discovery, root cause determination and fixes that ensue. PEs are called upon to collect and present data related to the issues over range temperature, process corner and voltage. These efforts are in addition to the other work required to test, characterize and ship early samples to critical customers.

The second part of the product life cycle is spent making the product more profitable (i.e., cost reduction) while ramping up production and resolving related yield and manufacturing issues. Later on in this phase the product is maintained and eventually, when time comes, led through obsolescence.

As a product engineer you are involved in all phases from day one! Once the first silicon is received you will "own the product". Basically, you are responsible for it from birth to grave. A very good analogy is raising children. And believe me I have lots of personal experience in that! You must spend a lot of time, effort and love in your child's early years. As a Product Engineer (or parent of this new "child") you must get involved early in the concept, definition and development phase in order to ensure the product will have minimum die size, highest possible test and fault coverage and debug capabilities in order to facilitate quick failure analysis and failure isolation in critical stages of life

**Supplementary Information** The online version contains supplementary material available at https://doi.org/10.1007/978-3-031-18030-9_1.

1
F. Barman, *Semiconductor Product Engineering, Quality and Operations*,
Synthesis Lectures on Engineering, Science, and Technology,
https://doi.org/10.1007/978-3-031-18030-9_1

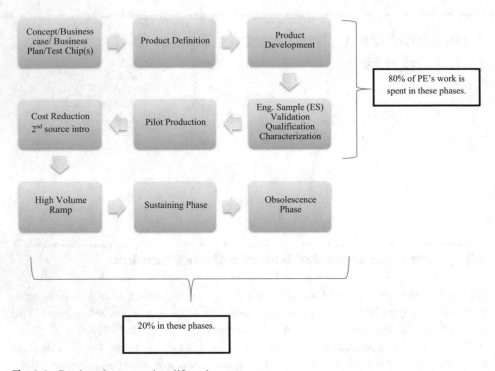

**Fig. 1.1**  Semiconductor product lifecycle

cycle. I will talk more about this in later chapters. The top half of Fig. 1.1 is where most of your efforts will be spent. If you have done your job well during these phases your work will dramatically decrease in the bottom portion.

Focus on technologies and methods that minimize die size. Die size will greatly influence the final product cost. Also, everything being equal, a large die will produce lower yield so the cost of each die will increase.

Another important area that PEs can have influence in is what DFT features are incorporated into chip design. PEs should be advocating for all basic DFT features available to the design team. On top of that, PEs should encourage design teams to incorporate DFT and observability features into custom designed blocks such as PLLs, DSP blocks and other custom designed sub blocks.

The next area that PEs must focus on is what capabilities can be incorporated into the design to facilitate all portions of the chip be exercised during burn-in operation mode. This is a critical feature to ensure that all major blocks are exercised during burn-in operation. If all major blocks are exercised during burn-in operations you can be assured that any faults, marginalities or shortcomings can be exposed during HTOL.

PE and operations should provide inputs early on related to memory testing (MBIST) schemes and capabilities, number of SCAN chains, PLL BIST, Analog debug capabilities,

DAC Signal to Noise Ratio, custom logic failure isolation capabilities. Understanding the whole chip and its capabilities and shortcomings is critical to future PE and Operations success in the new product introduction (NPI) activities.

Ask yourself, what PE and Operations group would like to see in test chip designs to de-risk any future failure analysis capabilities? Answers that come to you will be your guide.

Input on what "debug" features should be included in the chip design to address future failures during critical reliability qualification and bring-up activities.

It is very possible that due to workload for existing products PE will not be involved in the test chip stage. In that case PE should take time to provide inputs to design team leader and keep tabs on progress of the test chip once it is in validation and characterization stages.

At this point I would like to introduce the concept of "PE Hub of Influence". The PE is the hub of activity that keeps together the wheel that will turn and turn until ZDD421A is released into volume production (Fig. 1.2).

The product engineer must understand how each spoke in this wheel relates and works with other parts to enable a smooth operation through the product life cycle.

The Engineering team will need input from the PE to determine what test and quality features must be integrated into the product to enable reliability qualification, yield improvement and failure isolation during the product life cycle.

The sales and marketing team will ask the PE to provide a schedule for reliability qualification, characterization and other critical tasks. That team will also ask for cost estimates, yields and off-shore manufacturing capabilities for volume ramp phase.

The customers will ask the PE to provide characterization reports, specialized test reports relating to critical features, among many other requests.

The quality team will rely on the PE to understand how the product functions and how it will be operated during burn-in and other reliability stress tests.

The foundry team, which is the direct interface to the wafer fabrication vendor, will ask the PE to recommend where the fab process should be parked during volume production based on proper analysis of data collected from skew lots and product characterization.

The test team will look to the PE for input on how to organize the test program content to minimize test time. PE will also ask for certain data collection and specific format so yield analysis can be easily performed in off the shelf yield analysis tools.

Next I will expand on product engineer's roles and responsibilities during each stage of the life product cycle chart.

**Fig. 1.2** PE hub of influence

## 1.2    Concept/Business Case/Test Chip(s)

The marketing team has put forward a business case for ZDD421, a chip that must be reviewed and approved by the company management and Board of Directors. It will also need to be funded and resourced by the same team.

The business case includes chip architecture, big picture specifications, target markets, and target customers among many other details. Cost of product development will be greatly scrutinized and weighed against other options available to develop the product. Yearly and lifetime revenue of ZDD421A will be detailed in the business case. A proposed development schedule with major milestones will have to be put forward showing it will meet lead customer's development schedule.

As a rule of thumb lifetime revenue for a product should be between $10 \times$ and $20 \times$ the total development cost to be a viable product on which to spend company resources

**Fig. 1.3**  Test chip plan

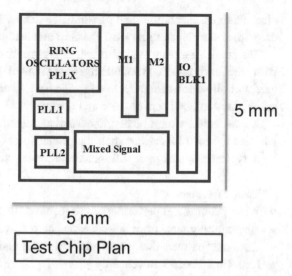

and time. For example, if a product costs $10 M to develop it should have the potential to bring in lifetime revenue of $100 M to $200 M. Of course there are exceptions to this guideline and that's where the marketing team will have to propose a lower bar and justify why it should be lowered. There could be strategic reasons or a long term engagement with follow-on products that may justify moving forward with development even though lifetime revenue is below $10 \times$ development cost.

One year before actual tape-out of, while the business case is being developed, the design team will have to tape-out a test chip intended to validate and characterize the more advanced and risk prone circuits that will eventually be incorporated into ZDD421A.

Critical designs included in this test chip are the special PLLs, mixed signal blocks, high speed IO interfaces and custom designed logic and memory blocks (Fig. 1.3).

This test chip will be critical to validate the analog block and a special PLL that is developed to enable high performance requirements of ZDD421A.

These types of small test chips are used by many fabless chip companies so wafer foundry will stitch many of these into a much bigger test chip. As a result, each wafer will contain test chips from multiple fabless companies. Of course on company A wafers only company A test chip is completed and accessible.

The test chip will be fabricated on the same process node as the final product. It will then be packaged into units and delivered to HQ so the design team can start validation and characterization activities for the critical blocks. The design team can also assess whether their design methodologies will create a product that has highest performance at lowest cost possible.

During this stage of product development the marketing team is focused on gathering inputs from lead customers on high level chip functionality, blocks, IO standards and how it fits into the system being built to utilize the full potential of ZDD421A. Also, what other features, capabilities and performance standards will the chip provide. A block level diagram is developed to kick off the chip architecture planning.

The marketing team will also gather competitive analysis of competing chip makers with similar products. The intelligence gathered from the competitive analysis exercise will allow the design team to understand where ZDD421A will stand compared to competitors in cost, performance and quality.

A business case must be developed and presented so the management team can evaluate whether to move forward or shelf the product pending a more solid case.

A typical business plan will include following main items:

- Lead customers
- What features and performance are required for lead customer to adopt this product
- Any commitments from these customers to develop products around ZDD421A
- Customer partners that would ensure success and their commitment
- Life Time Product Revenue by year
- Total Product Development cost
- Draft Development schedule

  o Dates for Tape out/first silicon/samples delivery/characterization/qualification /production release would be of great interest to product engineers.

- Blocks or IP that must be licensed or purchased from outside vendors and what's the cost.

Lead customers will require a proposed development schedule and what the budgetary pricing of ZDD421A will be. This schedule must meet the lead customer's system development schedule. If it does not they will push for an earlier date for engineering sample delivery. The sales/marketing team will be hard pressed to not agree to that request. Major milestone dates lead customer is interested in are:

1. Final feature review and lockdown.
2. Development review date for (RTL, synthesis, place and route, post layout timing, ECOs, freeze)
3. Tape-out date
4. First wafers out of fabrication
5. Engineering sample availability
6. Pre-production sample availability
7. Characterization complete
8. Product qualification complete should qualification be in caps?
9. Production volume shipment start.

### 1.2.1   Prepare a Budget for Key Items to Enable a Successful New Product Introduction

#### 1.2.1.1   Wafer Test

Planning and budgeting for wafer test equipment usually referred to as probe-card needs a few inputs to produce a reasonable estimate.

Type of probe-card has a big impact in it's total cost. So called direct dock cards where the ATE test head is directly docked to wafer prober via probe-card is the most expensive and complicated type of card that can be used. There are two main advantages to direct docking solution. We will explore these further here.

Minimal parasitic impedances due to direct contact between probecard and the die.

Use of the package substrate to convert the die configuration to the package configuration enables the same PCB to be used for both wafer test and package test. Therefore there is no need for a separate PCB design and manufacturing for package test.

2 Probecards @ $50,000 each. direct dock 1,200 Pins. $100,000

#### 1.2.1.2   Package Test

Package test presents a different set of challenges. The PCB design engineer must use the shortest distances possible between package solder balls and tester pogo pin connection points. This is necessary to minimize any additional parasitic capacitances and inductances these traces introduce. It is also very likely that here will be active and passive components on the package test board. These components are critical for performance, functional and IO leakage tests among many other uses. Majority of these components are capacitors that are used to smooth power up and down transitions as well as dampen and eliminate noise introduced during device operation.

- 4 DUT boards for package M + 4 DUT boards for package T. total 8 boards. $10,000/board. $80,000
- 2 sites per board total 16 sockets to fully populate all 8 boards. $2,000 per socket. $32,000
- 4 back-up sockets @ $2,000 per. $8,000

For package test PE should budget for $4 \pm 2$ DUT boards for each package type to accommodate early silicon development onshore and offshore transfer following close in its heels. Depending on the complexity of ZDD421 products we will need 4 boards per package (2) for a total of 8 DUT boards. Each DUT board is close to $10,000 in total cost.

Each DUT board could need between 1 and 8 sockets to have hooks in for parallel multisite testing in production. Some products start with single site and later if volume predictions come true they will embark on multisite testing DUT boards. Each socket costs between $1500 and $5000 depending on the product pin count.

### 1.2.1.3 System Level Test

- 4 boards + temperature control + PC etc. costs 4x $10,000 is total of $40,000

### 1.2.1.4 Qualification Tests

HTOL boards are used to exercise and stress the packaged units with accelerated temperature and voltage. These boards have to be rugged enough to stand 2000 h at elevated temperatures 125–150 °C. This will increase the complexity and cost of these boards to 2 or 3 times average PCBs. In addition each of these boards can run anywhere from 16 to 32 units in parallel. This will significantly drive the up the cost of these boards due to the specialized type of sockets that can withstand 2000 h+ at elevated temperatures.

- 16 units per board. 17 boards (2 AS BACKUP), total of 240 units can be stressed in parallel.

Need 3 lots to run in parallel. $3 \times 80$ units/lot = 240 positions total.
Cost/board $15,000. Total $150,000.

- HAST board: 10 units per board. 10 boards, each board $5,000. Total $50,000

HAST (Highly Accelerated Stress Test) boards have to be designed and manufactured using material that can withstand elevated temperatures and humidity (85c and 85% RH) for 100s of hours. As compared to HTOL boards these boards are less complex because they don't require the packaged units being exercised with inputs and clock while the stress test is running. Therefore these boards are less costly.

- L2 qualification boards: $35,000

This type of qualification test stresses the interface between packaged units and system boards that they will eventually be attached to. There is no voltage bias, clock or input applied to the unit during this test. The PCB they are attached to does what's called a daisy chain of a few strategically located product pins and can monitor the resistance of each daisy chain while the test is running. Depending on the size of the package units, anywhere from 5 to 15 units are mounted on a 16 layer PCB. Once the boards are populated with the packaged units, they are loaded in an oven that applies temperature cycling cycles that emulate the board assembly process.

- Biased and un-biased Temp Cycle boards. $15,000

As apparent from the title this test applies 100s of temperature cycles (cold □ Hot □ Cold....) to packaged units. There is no board needed if the test is not performed in biased condition.

Biased version of this test will require a board similar to HAST test in design and manufacture.


## 1.2.2   Failure Analysis Setup for $15,000

This type of hardware can vary wildly as you can imagine. Basically you will design a board that can access certain power and IO pins so the failing device can be analyzed in more detail. The design will have to include a way to observe the silicon chip inside from the top or bottom. With this capability while the device is running, internal nodes or areas can be observed for unusual behavior.

### 1.2.2.1   Software

- Yield Database and Analysis Software

Yield analysis and monitoring software is a well-developed field with many suppliers providing solutions for gathering, organizing, storing and analyzing data from wafer test all the way to package test and QA stages. These providers in general require an annual per license (1 license/1 engineer) and data storage fees. In general the faster and easier to use solutions cost more. A reasonable budget: $5000/year/license + $25,000/year of storage fees.

### 1.2.2.2   Statistical Analysis Tools

Wide range of products are available for statistical analysis ranging from Excel to RS-1. It is wise to budget for one license of RS1. It has decent capability for analysis and charting and is not a huge expense. Mybe around $2 K to $3 k.

### 1.2.2.3   Scripts for Specialized Analysis

Once you start collecting performance and other data from 100s of wafers and thousands of units it becomes necessary to create specialized scripts to extract valuable information and solutions from it. Hiring an intern or contractor to perform this maybe a valuable expense as it might reveal information that is not readily apparent from looking at raw data. It's basically a data science exercise for a very specialized set of data you have at your disposal. Budget $25,000/year for a part time person to perform this.

### 1.2.3 Outside Services

#### 1.2.3.1 Reliability Tests Execution
HTOL 0–2000 h (24, 48, 96, 168, 500, 1000, 1500 and 2000 h readouts) Total: $50,000.
ESD/Latch-up: $15,000.
Temp. Cycling: $25,000.
Tester time at outside test house:

- 6 months, 20 engineering hours per week @ $100/hour: Total of $48,000
- 3 months, volume testing @ $80/hour, 30 h/week: Total of $28,800

#### 1.2.3.2 Manpower
You will need manpower to test early ES units for distribution to internal teams as well as early shipments to customers. These test operators will also collect data for characterization and debug efforts. You will need technicians to bring up setups and make sure they produce repeatable results. Any equipment problems will need to be looked at and addressed by your technicians.

- Test Operators: 3 full time: 1 day time + 2 night shift
- Technicians: 2 full time: 1 day time + 1 night shift

Space considerations.

Consider the square footage needed to deploy test equipment including any ATE or system level test equipment.

Leave some room for unexpected equipment.

Label Makers are necessary so products can be labeled properly with part number, lot number, date code etc. before.

## 1.3    Product Definition Stage

During this phase PE should be pushing for full-chip burn-in mode to be able to fully assess quality and reliability of every block and the whole chip. At the least, the design team should maximize burn-in mode coverage as much as possible within the total chip size budget and other considerations. Some of these DFT features will be more easily integrated into the chip such as standard logic and memory blocks. Other blocks such as custom logic and analog blocks are more challenging to integrate DFT features. Nevertheless they must have some basic features so failure isolation can be achieved with reasonable efforts.

Other ideas for increased fault and failure isolation coverage are looked at in some detail below.

PLL bypass mode for debug and burn-in operation.

This feature enables the chip to be clocked by an external source rather than the internal PLL. This could potentially isolate any PLL issues during product rollout. There are limitations to this mode as there is much higher jitter and inaccuracy with an external clock source. If the design includes multiple PLLs then ideally each PLL would have a bypass mode Memory BIST is a MUST as today's chips utilize very large amounts of embedded memory.

Fault coverage is one of the most critical areas for future product success. Leading edge products have large blocks of embedded memory for a variety of features. PE must ask for memory BIST features to be implemented on every memory block whether small or large as a percentage of total chip size. Another useful feature to be requested is bit mapping capabilities for all memory blocks. This feature enables a PE to physically isolate failing bit(s) from logical addresses that are produced by basic memory test patterns.

Another important consideration for PE must be to know product operating temperature range. The wider the operating temperature range is, the more challenging and compli-cated the NPI activities become. Commercial grade is the most common with a range of 0–85 °C. Industrial grade range is a bit wider at −10 to 125 °C. Automotive grade cus-tomers mostly require −40 to 125 °C operating temperature range. Military grade products almost always require −55 to 125 °C operating temperature range. Completely different sets of equipment are needed to characterize and productize these different operating temperature ranges.

Operating temperature range that addresses the end market should be determined early in this phase to ensure chip design is executed with this as a requirement. Furthermore, design simulations must be performed to meet design specifications in this temperature range.

The PE should insist that market (s) segments addressed by the product are identified in this stage and the product is designed to meet specifications over the "superset" of temperature ranges.

The PE will Gain a very detailed knowledge of critical blocks and ensure adequate fault coverage and fault isolation capabilities are designed in. This strategy will increase the chances of isolating failures and fixing those issues in final design revision.

Once the PE has gained access to an early version of the datasheet, study the whole document but especially the Operating Condition section. This insures the PE is able to provide feedback to the development team on any challenges or obstacles you can predict and its impact on overall schedule.

Product engineers must push for memory failure bit mapping and physical isolation capabilities for all memory blocks.

There are a multitude of resources to learn about different algorithms used for memory testing and what advantages and disadvantages each one provides. With a few hours of focused research a Product engineer can learn quite a bit about this industry standard

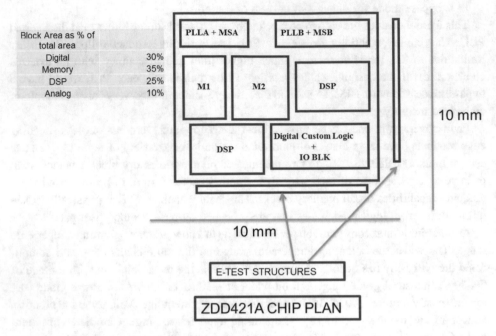

**Fig. 1.4**  ZDD421A Block Diagram MS: Mixed Signal Block

methods and likely contribute with good ideas to the design team. Rewrite last section of this sentence.

In the logic blocks, product engineers and operations team must push for an increased number of scan chains increasing fault isolation success rate (Fig. 1.4).

### 1.3.1  PE DFT Features Requests

#### 1.3.1.1  Acquiring Product Knowledge and Datasheet Study

Obviously, knowing the product in some details will go a long way in ensuring your success in resolving product issues quickly. Here are some ideas on how to approach this learning and optimize your time spent on this effort.

Review functional and performance specifications for each block and corresponding test plan for that block and gain a good understanding of how it operates and how it interfaces with other blocks. This is a study intensive activity as it involves reading the block description and discussing the details with design and test engineers to gain a full understanding and use that knowledge to make your work more productive.

Study the datasheet for ZDD421A and note functionality, critical performance criteria, IO specifications and power. All of these attributes will be characterized and validated once 1st silicon arrives. You will have a head start in resolving issues and analyzing data if you have already studied the datasheet and are familiar with it.

Helpful hints: Assuming there is more than one product engineer working on ZDD421A, assign each block to a particular PE and task him/her to study and then present their understanding of the block to the team. This method pushes the responsible PE to really understand the concepts, put that understanding on paper and be ready to present it to the larger team. Questions will be asked by more experienced members of the team enabling better understanding and bringing fresh thinking to the table. This must be done over several sessions until all blocks are covered and well understood. A typical presentation will address the following questions:

- Other areas to explore in gaining better product knowledge:
- How does the block fit into the bigger chip?
- What are the most important functions of this block?
- How does it affect and interface with other blocks?
- What is the test plan for the block? What will be the estimated fault coverage for this block?
- Where are the test coverage shortfalls? How do we address them or have a workaround?
- Is this block designed internally or licensed from a third party?
- What is the simulated fault coverage for the block?
- Does it meet internal standards for fault coverage?
- What is the path to fault isolate future failures that may occur in this block?

Look carefully at Absolute Maximum Operating Conditions as you will be tasked with validating that ZDD421A will meet the requirements outlined in this section of the datasheet. Decide which test platform, ATE or System Level, you will use for validation.

Studying the Operating Conditions sections of the datasheet will show you what temperature range(s) your characterization and reliability qualifications plans will have to cover. This section will also outline the operating voltage conditions the validation, characterization and reliability qualification will have to cover.

Commercial temperature grade products cover 0 to 85c and are the most straightforward to characterize and qualify.

Industrial speed grade products are more challenging as they cover −25 to 110 °C. This involves equipment that can go down to −25 °C which is somewhat more challenging due to moisture in the environment near the unit under test. This range is manageable with off the shelf equipment that can force −25 °C with pretty good accuracy and ease.

Automotive grade products often require −40 °C in the lower end which is higher complexity to execute and maintain. Equipment that can guarantee this temperature is $2 \times$ to $3 \times$ more expensive than commercial and industrial grade equipment. These category of products also require testing and guarantee of performance at 125 °C. The hot temperature condition can increase power consumption as compared to commercial operating conditions for the same product. Also operating a product at 125 °C will decrease it's life time and require more margins and guard bands to be added to test conditions such as voltages, timing and temperature.

Military grade products require −55 to 150 °C and are the most challenging products to characterize and qualify. These types of products will require additional reliability qualifications tests to ensure that they can withstand harsh conditions in battlefield or space (such as radiation in space).

Automotive and Military grade products must conform to respective industry standards such as AEC100 or MIL883 and other applicable specifications before they are considered as qualified for production. Any subcontractors you use to manufacture your products must be certified to comply with automotive or military industry specifications that govern quality of these products.

Look out for products with multiple core and IO voltages. These types of designs require robust level shifters designs between different voltage domains. Designs with multitude of IO voltages require a few IO designs working over an extended range which pushes some supplies into ranges that are nearing over stress and possible failure. Other specifications to pay attention to are the VIH/VIL and VOL/VOH which will have to be characterized. What range of current loads the IOs are specified to handle will influence the decision as to which ATE to use?

Review key timing diagrams and become familiar with input and output timings.

Get a hold of the complete product pin list in a spreadsheet format as it will be used extensively for ESD and latch-up tests during reliability qualification. This information will also be needed to design the burn-in boards and other hardware.

Product package outline drawing which shows the physical dimension of each unit is also very important to understand as it will be required for all hardware designs inducing test, characterization and reliability qualification.

During reliability qualification, manufacturing tests and customer validation failures will inevitably occur. You must have capability to identify the failing block(s) for critical failures and quickly identify the location of the failures to within a few transistors to enable any physical failure analysis. That is a very high bar to set!! However, if you do not put a stake in the ground that "fault isolation per block is essential to success" you will end up with a product that is difficult to qualify and move into a volume ramp.

## 1.3.2  Critical DFT Features for ZDD421A

### 1.3.2.1  PLL Bypass Mode

This is an especially useful tool for burn-in operation, fault isolation. In this mode you will be able to provide clock to the chip from an external source, basically "bypassing internal PLL". Burn-In chambers in general do not have capability for high speed clocks and this feature will be critical to getting the product into proper burn-in mode using low speed clock from the chamber. A good rule of thumb is a 5 MHz clock being the top speed a burn-in chamber can provide. PLL bypass mode is also very useful in analyzing certain class of failures that might be related to clock skew or jitter.

### 1.3.2.2   BIST Insertion for Critical Blocks

Memory BIST bitmap tool capability from logical to physical address translation capability is a must. Memory failures must be isolated to a particular bit so detailed failure analysis can be performed. If memory blocks are licensed from an outside vendor, bit mapping and failure isolation capabilities must be well understood and assessed before finalizing the agreement to purchase such IP. In the ES phase one of the activities that should be on the checklist is to find a memory failure and use it to pipe clean the failure isolation flow. This activity involves taking a failure from electrical identification all the way to physical location identification of the failing bit.

SCAN ATPG insertion is a must with fault coverage goals generally in the 98% + range to meet reasonable quality goals.

TDF (Transition Delay Fault) test patterns are also a must have to exercise many paths and blocks at high speed to uncover any delay faults. These types of test patterns have lower fault coverage as compared to SCAN ATPG because they require the design paths to meet high speed timing requirements. Fault coverage goals for these types of test patterns are in the mid to high 80s% range.

Custom designed logic or DSP blocks should have fault isolation and performance metering capabilities built in to enable critical speed path analysis and basic performance measurement capabilities.

JTAG access to critical blocks will be needed to ensure in burn-in mode the internal chip blocks can be accessed and exercised using JTAG.

### 1.3.2.3   Burn-In Mode is Critical

Without a robust BI mode you will be hard pressed to evaluate the reliability of your product.

Most critical features of BI mode:

i.   Ability to toggle all blocks either sequentially or in parallel through a low speed external clock?
ii.  BI mode must have the ability to load SCAN chains with data and propagate it through logic. It is not necessary to capture outputs. But there should be capability to ascertain that data propagation has been successful.
iii. Memory BIST functioning and capturing output for all memory blocks.
iv.  Voltage and temperature acceleration capabilities are MUSTS.

v.  Ability to raise core voltage by $+10$–$20\%$ while blocks are in functional or BIST mode.

vi.  Raise IO voltages by as much as $10$–$20\%$ above nominal during BI while the product is in functional mode.

### 1.3.2.4  Very Low Power Mode for Biased HAST

Biased HAST test requires the product to have a "very low power mode" to meet stress test requirements specified in JESDZ22-A118.

A DC voltage bias is applied to all Vcc and IO voltages during this test but the unit under test must maintain a very low power mode so operation does not affect the $85\%$ humidity saturation required around the DUT. Basically if the device under stress consumes too much power, the moisture that is supposed to penetrate the product will not be able to do that.

Test Conditions:

110 °C/85%RH, 130 °C/85%RH

HELPFUL HINT: This may be a challenging goal so it's important to make this request early on to give the design team time to react and implement.

### 1.3.2.5  Tetramax License and Capability to Do Fault Isolation is a MUST

Tetramax is an essential tool for isolation and analysis of SCAN ATPG and TDF failures. PE must ask for this tool to be acquired and installed. Tetramax requires a license that is 10 s of thousands of dollars.

HELPFUL HINT: put this in your budget and request early.

Capability to process SCAN ATPG failure logs through Tetramax and do fault isolation with high confidence level.

Failure pin(s) and timing cycles must be recorded for each failure in a particular format so that it can be processed through Tetramax tool. Enough pattern cycles must be recorded before and after the failing cycle so analysis process produces a high confidence in location of the failure.

## 1.4    Product Development Stage

Following the product definition stage where features and specification are locked down, the product development stage presents its own challenges for product engineers.

During this stage the design team may drop certain requests in order to expedite development schedules and facilitate design implementation. PEs must be on the lookout for these situations and push the design team to keep the important features or have a good reason why it should be dropped.

Engage your upper management early on so that they can provide needed support in pushing for implementation of critical PE features.

Once the preliminary block sizes, total chip size, packaging and test information have been determined, the Operations team should start developing early yield and cost models to prepare for such requests with ease. Getting involved in this activity will also enhance PEs understanding of critical blocks and how they will affect future yield and characterization procedures.

Any additional or unexpected costs associated with complying with Reliability Qualification activities should be introduced and discussed early on. This concept will be introduced in Chap. 3. Based on my experience these costs mostly consist of production burn-in, sample burn-in or other monitoring schemes like ongoing reliability monitoring. In addition to ATE testing, a system level test may be required for some time to ensure an acceptable outgoing quality level for ZDD421A.

Operations team must provide input on packaging options and any drawbacks with certain technologies. Product package options will greatly influence future failure analysis capabilities. Once the silicon die is encased in the packaging it becomes much more difficult to reach the silicon die to isolate failures or do detailed failure analysis. All the features PE requested in the definition stage will be useless if they are not accessible in the packaged unit. PE must push for this accessibility in all package options for ZDD421A. Amongst those pins are PLL output observation, scan chain in/out, memory failure information pins. Not sure what pins means.

The wider the operating temperature range the higher the final product cost will be. Equipment that test or handle chips from $-55$ to $125\,°C$ (so called HANDLERS) is $2 \times$ to $3 \times$ the cost of commercial temperature range equipment. These types of equipment are also more difficult to operate and maintain. They require liquid nitrogen cylinders to bring the temperature down to $-40$ to $-55\,°C$. Testing silicon in wafer form is also much more complicated and costly at temperature ranges beyond commercial grade. Wider operating temperature range also affects voltage and temperature acceleration settings during HTOL!

In advanced silicon technology nodes transistor performance degrades over time due to many different factors that are out of the scope of this book. This phenomenon is commonly referred to as "aging". OPS team must request a design team to perform aging simulations based on models provided by wafer foundry and present results confirming that after operating for 10 years ZDD421A at nominal conditions ZDD421A will meet datasheet specifications. This is a general rule of thumb for reliability purposes. A specific product might only need to meet 5 years or even less, it depends on the market segment being addressed. This will put pressure on design team to build in enough design margins to compensate for performance degradations as result of aging.

In every silicon process node there are multiple types of transistors that are available to chip designers:. P-channel, N-channel, Low Vth, High Vth, thick gate oxide and a few other types. PE must understand the make-up of major blocks by transistor type. This information will determine which transistor type parameters should be focused on during first silicon bring up if performance is not meeting datasheet specification or other issues pop up.

The engineering team must also perform Electro-thermal simulation to determine if more heat dissipation capability should be added to ZDD421A. There are off the shelf solutions available for performing these simulations. This is a must as it will affect reliability of the product. If ZDD421A cannot adequately dissipate power during normal operations, certain units will go into thermal runaway and damage the component and/or a complete system board. This analysis will also be influenced by what is the form factor of the system that will house our chip. What's the airflow available during normal system operations? This analysis will also need as an input, typical and maximum total power dissipation.

Start putting together a detailed yield model for ZDD421A. This will guide you on whether ZDD421A is healthy enough to be released to volume manufacturing. This model should be broken down by major blocks and mirror what will be in the manufacturing test program. We will discuss this topic in more detail in Chap. 4.

Another area to focus on during this phase of the ZDD421A is the flow and structure of the test program you wish to see during engineering sample (ES), reliability qualification, characterization, early production and volume ramp. I will present more details on this topic in Chaps. 3, 4 and 5.

## 1.5   First Silicon, Product Validation, Product Qualification and Characterization Phases

This is one of the most intense phases of the product life cycle chart. All three activities are happening at the same time and each one requires product engineer's attention and time.

Basically in this phase, product engineers will receive first units assembled from the very first wafer rolling out of fabrication facilities. These units will need to be received, and visually inspected to ensure there are no gross defects that would prevent validation start. Any minor defects or issues observed should be recorded so they can be addressed later. Following visual inspection the units will need to be tested with a basic test program and quickly distributed to the engineering team so they can start validation and other activities. Units must also be distributed to Test Engineers so they can start validating and building the test programs required in later stages of the life cycle. Of great interest in this time period is the engineering sample test programs that will be used to ship the first units to lead customers. Distribute the first sets of units to internal groups ASAP.

A checklist of items that need to be tracked and completed is a good place to start organizing activities for the ES phase. This checklist will also enable assignment of critical resources and equipment to tasks on an individual level. This will be critical in managing activities and ensuring that they will be completed on schedule. In Table 1.1 I have provided an example that can be easily modified to meet project requirements for different products.

Characterization of skew lots will start in this phase and will provide critical information to all teams as to whether ZDD421A meets functionality and performance goals and with what type of margins. This activity will also provide the information required to finalize the product datasheet and publish to customers (Table 1.2).

These are some of the tasks a product engineer should pay attention to in this phase.

Deliver the first set of units to internal groups.

Work to get an early test program that can be used for LU and ESD testing.

Work with TE to release an ES test program with "best test coverage possible." This is critical to meeting early customer sample demand quantities.

Work with material planning team to start reliability qualification lots (and backup lots) so you don't lose any time waiting for this material.

Receive reliability qualification test program and start executing your plan.

Receive characterization test program and start executing the plan.

Further develop your manufacturing flow validation plan based on experiences gained during this phase.

During this phase it may be determined that there are issues that can only be addressed by changes in certain mask layers. If this decision is made it means a large sum will have to be spent to change, validate and remake those mask layers. This is called revision B tape-out. This is a large effort but is quite common in the semiconductor industry. The work required to release version B to volume manufacturing is mostly incremental to what was done in revision A. Assuming there are no major changes to the design, data collected on revision can be supplemented with additional data from revision B to complete the work.

## 1.6 Pilot Production Stage

At this stage of the product life cycle the samples of final revision of ZDD421A design have been delivered to lead customers and these customers have validated the changes in revision B and it meets all their requirements for performance, quality and cost. The engineering team has validated all changes and there are no more changes required before going into pilot production stage. Characterization and reliability qualification activities have been successfully completed and we are ready to manufacture the first few production batches and deliver units from these lots to a wider pool of customers. Monitoring the progress of these lots in the fabrication facility is critical to make sure any issues or

**Table 1.1** Proposed budget for hardware, software and outside services to achieve NPI goals

| Item (hardware and software) | Quantity | Cost | Total |
|---|---|---|---|
| Wafer test probe cards | 3 | $50,000 | $150,000 |
| Package test | 4 | $10,000 | $40,000 |
| Sockets | 8 | $2,000 | $16,000 |
| System level test HW | 4 | $10,000 | $40,000 |
| HTOL board | 17 | $15,000 | $255,000 |
| HAST boards | 10 | $5,000 | $50,000 |
| L2 qualification boards | 5 | $3,000 | $15,000 |
| Biased HAST and temperature cycling | 5 | $5,000 | $25,000 |
| Failure analysis hardware | 2 | $5,000 | $10,000 |
| Yield analysis software | 1 | $75,000 | $75,000 |
| Statistical analysis software | 1 | $3,000 | $3,000 |
| Specialized scripts for data analysis | 1 | $15,000 | $15,000 |
| Total | | | $694,000 |
| Outside services | Quantity | Cost | Total |
| HTOL execution | 2000 | $ 15/h | $30,000 |
| ESD/latch-up execution | 5 | $3,000 | $15,000 |
| Temperature cycling | 1 | $20,000 | $20,000 |
| Engineering tester time (5 > test house | 480 | $100 | $48,000 |
| Volume testing @ test house | 360 | $80 | $28,800 |
| Product holding trays | 2000 | $2 | $4000 |
| Bake oven | 1 | $5,000 | $5000 |
| ESD safe bags | 1000 | $2 | $2000 |
| Moisture barrier bags | 2000 | $2 | $4000 |
| Vacuum sealer | 1 | $3,000 | $3000 |
| Strapping machine | 1 | $2,000 | $2000 |
| Strapping material | 2 | $250 | $500 |
| Shoe boxes with logo | 2000 | $1 | $2000 |
| Total | | | $164,300 |

delays observed are addressed quickly and adequately. These first few batches must be tested at wafer level and yields compared with revision A to make certain that the changes made to revision A are having a positive effect on yield. These wafers will have to be assembled and then tested at package level so the product engineer can make certain yield at this test step meets or exceeds expectations.

**Table 1.2**  New Product Checklist

| Item | How many | Who | When | Notes |
|------|----------|-----|------|-------|
| Blind build wafer selection/decision | | | | |
| Blind build heads up to assembly subcontractor (qty, package etc.) | | | | |
| Blind build engineering assembly instructions to assembly subcontractor | | | | |
| 3 Day FASTRACK | | | | |
| Internal checkout unit tracking, test and distribution | | | | |
| Spreadsheet to track units and test results | | | | |
| Ready for set-up of ZDD421A SLT stations | | | | |
| ATE tester availability | | | | |
| Probecard wafers | | | | |
| WS test program | | | | |
| WS initial bring up | | | | |
| WS TE responsibility | | | | |
| WS PE responsibility | | | | |
| Assembly | | | | |
| Test program checkout and release procedure and checklist | | | | |
| ZDD421A test plan | | | | |
| PE for FT | | | | |
| TE for FT | | | | |
| Loadboard | | | | |
| Sockets | | | | |
| Change kit | | | | |
| Probecard (SPECIAL SUBSTRATE FOR PROBECARD BUILD) | | | | |
| Socketed probecard | | | | |
| Substrates | | | | |
| WS traveller | | | | |
| Yield estimates/goals | | | | |
| 1st lot sort and test program and direct dock prober | | | | |
| Characterization plan and execution | | | | |
| Corner lot and units | | | | |

(continued)

**Table 1.2** (continued)

| Item | How many | Who | When | Notes |
|------|----------|-----|------|-------|
| Char. Test program | | | | |
| Rel and Eng. units build/test | | | | |
| BI pattern | | | | |
| BI board design inputs/progress | | | | |
| BI socket | | | | |
| Total BI checkout | | | | |
| 100 unit early BI plan, execution, FA, reporting | | | | |
| Final test program for BI time points | | | | |
| ESD/LU | | | | |
| Bias HAST (board, checkout etc.) | | | | |
| Temp. Cycle | | | | |
| HTS | | | | |
| Mechanical shock | | | | |
| Mechanical vibration | | | | |
| Curve trace board for engineering work | | | | |
| Sort 1st lot and ship for assembly | | | | |
| Sorted wafer build plan and engineering instructions for assembly to Amkor | | | | |
| Reserve prober | | | | |
| Package substrate package A | | | | |
| BOM | | | | |
| FT traveller | | | | |
| Loadboard | | | | |
| FT socket | | | | |
| FT test program | | | | |
| Secure ATE | | | | |
| SLT system package A | | | | |
| SLT socket | | | | |
| SLT test program | | | | |
| Secure SLT system | | | | |
| TEC for package A | | | | |
| TEC for package B | | | | |

(continued)

**Table 1.2** (continued)

| Item | How many | Who | When | Notes |
|------|----------|-----|------|-------|
| Package substrate Package A | | | | |
| BOM | | | | |
| FT traveler | | | | |
| Loadboard | | | | |
| FT socket | | | | |
| FT test program | | | | |
| Secure ATE | | | | |
| SLT system for package B | | | | |
| SLT socket | | | | |
| SLT test program | | | | |
| Secure SLT system | | | | |
| TEC Head | | | | |
| OPNs/MPNs set-up in manufacturing systems | | | | |
| Glalaxy databse set-up for ZDD421A | | | | |
| STDF naming convention | | | | |
| Using Galaxy for ZDD421A chacterization | | | | |
| How do we handle 1 die to many package units? | | | | |

At this stage the product engineer will execute the manufacturing validation plan which will be presented in Chap. 5. One of the main responsibilities of the product engineer is to identify any issues or roadblocks presented in this phase so ZDD421A can move into volume manufacturing. Another important activity is to ensure that the chosen yield management system (YMS) can receive wafer e-test, wafer test and package test data without any issues. Product engineers should also exercise the YMS with data from these lots to validate the process of loading data, storing, charting and analyzing the data is operational.

## 1.7 High Volume Ramp Phase

During this phase hundreds of wafers will be started and processed in the fabrication facility. The large amount of data produced from the testing of these wafers and resulting final products must flawlessly move into the yield management system (YMS) so yield analysis is possible. Regular yield and quality metrics reviews are a must in this phase as large numbers of wafers may be effected if yield or quality issues are not addressed early and quickly. Product engineers should also devise a plan to regularly sample production material and put that material through a selected list of reliability qualification stresses. These exercises will reveal any quality or reliability issues due to process changes or drift.

## 1.8    Cost Reduction Phase

In the cost reduction phase the focus will shift to reducing the number of manufacturing and test steps while maintaining quality and reliability. There is a certain balance that needs to be reached so at the end you deliver a product that is low cost and reliable. It is not an easy task. Also in this phase a second source for fabrication, test and assembly are looked at and selected if volume projections justify it. Increasing the product yield at every test step is critical in this phase. We will dive deeper into this topic in Chap. 5.

## 1.9    Sustaining Phase

By the time this phase is reached the product has been running for some time and a baseline, yield, quality and reliability measures have been set up. Any excursions from these baselines must be looked at and analyzed carefully by the product engineer. Any sustained yield or quality issues are of concern in this phase as it may be indication of changes or drift in the processes put in place to manufacture ZDD421A. Product engineers must monitor yields on a regular basis to ensure problems are identified and addressed quickly. YMS must be set up to alert the product engineer if yield or quality fall below a certain limit.

## 1.10    Obsolescence Phase

This phase of the product life cycle may not arrive for 5–10 years after new product introduction, sometimes even longer for certain products. In general, customers are made aware of this phase well ahead of time so they can place their last time buy orders and negotiate any extensions to the proposed obsolescence timeline. Product engineers and Operations team's role in this phase is limited to ensuring that test hardware, test programs and all the material required for finishing and shipping the product are available and stocked to meet customers' demand. Generally customers are allowed to take six months to evaluate their demand and other concerns they may have with obsolescence of a certain product. Following that there will be formal notification issued in which, a timeline for last time buy order placement and product delivery is proposed.

In this part of the product life cycle the revenue and the number of customers will decline significantly and the costs of sustaining the infrastructure and the resources needed to support the product become harder to justify. Precious resources like engineering talent and funds can be alternately deployed to new products which will bring in new customers and have the potential to create significant revenue growth.

Products in automotive and communications end markets have a very long period before they enter the obsolescence phase, sometimes taking up to 10 years to reach this phase. Consumer market oriented products will reach this phase sometimes as early as one year after volume ramp. Multi end market products may be declining in one customer while other customers maybe still in the high volume phase.

You may also think twice about starting the obsolescence phase of a product that is a strategic solution for one of your valued customers but has declining revenue in multiple other customers.

Some products may have to be obsoleted because they are not profitable anymore. This could be caused by several factors such as price drop or increased cost of manufacturing. Price erosion is generally due to competitors entering the marker with a similar product and offering the same customers a better deal! Approaches to save these categories of products are outside the scope of this book. Semiconductor fabrication facilities sometimes increase the price of older technologies as a way to move their customers to more advanced technologies. Also, older technologies generally require older equipment that may be getting more expensive to repair and maintain therefore increasing manufacturing cost.

Another strategy for pushing customers to move to new products is to gradually increase the price while offering newer products at a lower cost.

If you are considering obsoleting one of your products you should first compile a list of customers and their revenue for that product. Then you will need to review the strategic value of each customer at the moment and in the future and assign a value to it.

Once you have decided to move a product into the obsolescence phase you will need to work closely with marketing, sales and other groups to put together a presentation that will clearly explain to your customers why you are making this decision. They will not be happy about it but you can mitigate that by offering an obsolescence plan and schedule that will give them enough time to determine how to move forward.

Last time buy order placement date is critical since you will gather all the orders from various customers and then place a onetime bulk order with your suppliers of wafers, packages, testing etc. Once this step is completed you will have to limit any future orders being placed. There may be special cases where you might have to be flexible to accommodate a customer that didn't quite meet the deadline.

You will then work with your suppliers to get dates for when these orders will start hitting your finished goods inventory. Once you are comfortable with those dates you will need to make delivery date commitments to your customers.

In general the last time buy orders for a declining product will be much higher than the existing order rate so you will need to make sure that suppliers can provide enough wafers, packages and test capacity to meet the increased demand. If availability of these materials and services are limited then you will have to plan for a longer lead time between order placement and shipment of the product.

## 1.11  Summary

Semiconductor product life cycle expands multiple years and product engineer's roles and responsibilities evolve during this phase. 80% of the work is concentrated in the first one or two years of the product life cycle. Product engineers must be on the lookout for issues that may seem insignificant to other team members and raise red flags for critical issues that could affect yield and quality of the product.

During each phase of the product life cycle the product engineer and operations team has to plan and execute certain activities as outlined and detailed in future chapters.

# Reliability Qualification

**2**

## 2.1 Qualification Standards Pyramid

The Qualification Pyramid will serve as the structure to organize brings up, characterization, qualification and any other related activities during critical engineering sample phase (ES).

There will be lots of work involved in fully executing these complicated plans. Generally the Reliability team heads these activities with close cooperation with product and test engineers. These stress tests are the "obstacles" on the obstacle course and their purpose is to expose any weaknesses in ZDD421A before the product goes into production phase. The design team that is developing ZDD421A will work with other teams to understand the failures, root cause them and quickly put in place a fix to address the issue. This means that the stress test that exposed this weakness will have to be repeated with material that has the fixes incorporated in design and fabrication. Stress tests that pass with flying colors will not have to be redone unless there is a good reason to do so.

In general the flow of execution for a stress test is shown in Fig. 2.1

The test program at time zero must be the same as test point 1 and any subsequent testing that is performed. The test program has to stay constant so the product performance can be assessed fairly and any failures are attributed to the product going through that specific stress test.

Shift analysis is exactly what it sounds like! Product engineer is looking for any shifts in critical parameters of ZDD421A as stresses are accumulated during the qualification process. Some examples of these parameters are:

**Supplementary Information** The online version contains supplementary material available at https://doi.org/10.1007/978-3-031-18030-9_2.

F. Barman, *Semiconductor Product Engineering, Quality and Operations*,
Synthesis Lectures on Engineering, Science, and Technology,
https://doi.org/10.1007/978-3-031-18030-9_2

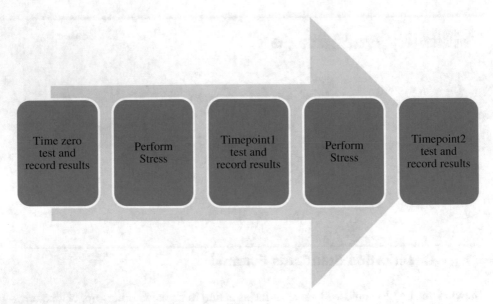

**Fig. 2.1**  Reliability qualification process cycle

- Minimum voltage a memory block can operate at.
- Maximum speed of a certain path or block in ZDD421A.
- Changes in currents such as leakage or power supply current capability.

That is where the product engineer adds value by using the knowledge gained during product study and applying that to come up with a list of critical parameters that should be part of shift analysis.

One of the outputs of this extensive activity is the final ZDD421A Reliability Qualification Report. This report is the main formal document that you will be presenting to customers to gain their approval and strengthen their confidence in ZDD421A. Reliability Qualification Report includes pass/fail test results as well as "major parameter shift analysis".

Majority of the product's customers will ask for this report among a few other items. However, I always advise PEs to thoroughly read "major customer" specifications to ensure your plans cover their requirements. If there are gaps, you will need to address these with your customer's quality manager and come to an agreement on how to close these gaps.

Figure 2.2 shows what I call the Reliability Qualification Pyramid. At the base of the pyramid is the product datasheet. That's the document that outlines product functionality, operating modes, operating conditions for temperature and voltage, power consumption for static and operating, timing diagrams, complete pin list and finally the form factor of the product with relevant dimensions and many other details. This document serves

**Fig. 2.2** Reliability
qualification pyramid

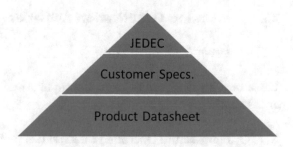

as the most efficient method to communicate to customers how ZDD421A operates and how to deploy it in their end application. Every specification in this document has to be validated and characterized. Functionality is validated either on ATE (Automatic Test Equipment) or a validation board which at it's center has ZDD421A. This board will have several interfaces to monitor functionality, power consumption and other vital signs of functionality like certain bus or pins of the ZDD421A. Use this document to create the different versions of the test program and voltage, temperature conditions for any reliability qualification tests.

Middle tier of the pyramid consists of documents or requests that customers provide to define what they would like to see for functionality, performance or reliability. This is a very important document to read and understand before planning any activities related to ZDD421A.

The top tier of the pyramid is JEDEC specifications. JEDEC is an organization that works with semiconductor industry leaders to develop and publish test and other specification standards for semiconductor based products. A great majority of this organization's specifications are free and available at www.jedec.org. JEDEC publishes hundreds of specifications for various types of environmental and stress tests. Each specification includes a very detailed description of how the test is to be performed, what environmental conditions must be present and how to evaluate whether ZDD421A passes or fails that particular stress. Many customers will directly refer to a JEDEC specification number in their qualification documents. I will refer to some of these documents later on in this chapter.

Customer specifications come in all different forms, however one thing in common is their specific requirements for test(s), certifications etc. that they would like to see done and pass on ZDD421A. Some of these are very detailed and some are very basic. You may want to contact a customer representative responsible for this product to answer any questions or seek advice on particular tests stated in their document. Smaller customers will likely accept your plans and reports. Bigger customers will have lots of requirements. In my experience you can negotiate to remove some items and satisfy other requirements with little to no changes to your plans.

The product's datasheet will define functionality, major blocks and key parameters for shift analysis as well as the basis of all test programs developed for ZDD421A.

## 2.2   Planning Qualification Activities

### 2.2.1   Level 1 Qualification

**These tests are designed to stress the die to expose any weaknesses in design, process or assembly.**

1. **High Temperature Operating Life Test (i.e. Burn-In)**
   a. JESD22-A108-D, T = 130 °C, Duration 1000 h. Some customers ask for 2000 h or as "information only".
   b. Typical sample size for this test is $77 + 5$ as safety in case there is an issue holding some units back for analysis. Intermediate read points are 24, 168, 500, 1000 and 2000 h.
   c. 3 different fabrication lots spaced a few days apart in start date with production POR process and centering.
   d. Purpose: Stress the units at accelerated voltage and temperature to simulate 10 years in ~1000 h!!!!! 10 years equals 87,600 h. That means you need at least a Total Acceleration Factor (TAF) of 87.6 (usually rounded to 100 for ease of calculations).
      i. TAF = Voltage AF × Temperature AF
   e. What do you need to execute this test?
      i. Final pin out and preliminary datasheet.
      ii. BI board design, schematic, layout and manufacturing.
      iii. Foundry reliability calculator for the particular process being used to manufacture your product.
      iv. Reference for more details on "calculating reliability".
      v. https://en.wikipedia.org/wiki/Reliability_(semiconductor)
      vi. https://www.intersil.com/content/dam/intersil/quality/rel/calculation_of_sem iconductor_failure_rates.pdf
      vii. https://www.sony-semicon.co.jp/products_en/quality/pdf/qr_chap2e_201604. pdf
      viii. What BI oven will be used to run HTOL?
      ix. Helpful hints: Target using readily available BI ovens that can be procured in US and off-shore, in case you need to do BI during the production phase.
         1. Using proprietary ovens will increase the cost of BI board and BI execution dramatically.
   f. BI oven considerations:
      i. There are many types and brands of BI ovens. Majority of ovens have capability to run patterns while devices are powered up and under stress conditions. Some have capability to individually monitor the units and ensure case temperatures are controlled avoiding thermal run-away or other adverse conditions.

    ii.  Some ovens only control the ambient temperature of the oven. These types of ovens are not suitable for high powered products.

   iii.  Most commercially available ovens have limited pattern drivable pins, clock speed (~5 MHz) and pattern memory depth.

   iv.  Local reliability vendors in the Bay Area provide services such as BI oven, board design/manufacturing and socket selection.

    v.  Burn-In board design for the product that meets the requirements of the oven chosen.

   vi.  How many voltage supplies does the product need?

   vii.  What power will be consumed by each power supply during high voltage/high temperature conditions while functional patterns are running?

  viii.  How many parts per board in the final design?

   ix.  BI Pattern that will exercise every block.

  g.  BI pattern checkout considerations.

    i.  Checkout the BI pattern on the ATE and make sure the pattern passes on a typical unit and all indications are that the patterns is exercising all blocks sequentially or in parallel if the capability exists.

    ii.  Also run the pattern at elevated voltage and temperature on the ATE to make sure it behaves as expected.

   iii.  Then the pattern has to then be translated to the BI oven format before it can be used in that environment.

   iv.  Load a few units onto to the BI board and load into the BI oven at room temperature and typical voltages.

    v.  Do the units behave as expected?

   vi.  Start increasing voltage and temperature and keep observing the unit behavior.

   vii.  Is it as expected?

  viii.  Resolve any issues that might be observed.

   ix.  Run the experimental units for a few hours at elevated voltage and temperature and then take them to ATE to see if they still pass.

  h.  Test Program Considerations:

    i. Nearly Full Production Test Program with following parameter search and log capability:

      1. Vmin searches for critical blocks. Memory, logic, Analog

      2. Analog parameters

      3. IO Leakage

      4. Power measurements.

2.  **ESD**

  a.  Typically 3 fab lots, 5 units/lot

  b.  Human Body Model, JESD22-A114F, $\pm 2$ kV

  c.  Charged Device Model, JESD22-C101F, $\pm 100$ V

  d.  Machine Model, JESD22-A115C, $\pm 100$ V (Not in use any longer)

    e. High Speed IO pins like SERDES, PCIE etc. will not withstand 2 kV HBM or
       200 v CDM

    f. Purpose: Stress ESD protection for power supplies and IOs.

    g. What do you need?

       i. IO groups (SERDES, GPIO),

       ii. Vdd groups (core, IOs)

      iii. Vss groups

      iv. IOs with NO ESD protection.

       v. High Speed IO pins like SERDES, PCIE etc. may not withstand 2 kV HBM or
         100v CDM

      vi. Usually a multi-tab Excel worksheet

3. **Latch up**

    a. Typically 3 fab lots, 5 units/lot

    b. JESD78D, 100 and 25 °C, $\pm 100$ mA

    c. Purpose: Determine if IOs or power supplies can withstand certain conditions and
      not latch-up.

    d. Same information used for the ESD test can be used here.

## 2.2.2  Helpful Hints

Finalized Device Pinlist is a basic requirement for all groups to work off of.

- Usually a multi-tab Excel worksheet.
- IO groups (SERDES, GPIO)
- Vdd groups (core, IOs)
- Vss groups
- IOs with NO ESD protection.

High Speed IO pins like SERDES, PCIE etc. may not withstand 2 kV HBM or 100v
CDM. Consult IO design manager.

    "Quick ESD and LU" stress ATE test program at a minimum requires the following:

- Power Supply Shorts Test.
- Continuity (Open/Shorts) Tests.
- IO Leakage Tests (H/L/Tri-State).
- Basic functionality test.

The following tests are targeted to stress the substrate/die/underfill combination and
interface to expose any weaknesses in structure or material.

Helpful hints:

It is very important to review the materials being used for assembly (BOM) of the product looking for places that the suppliers(s) may have used inferior materials to save cost. Once the BOM is finalized it must be filed into a document control system ender revision control and change sign-off loop.

This BOM is used for building package qualification lots.

Usually an "assembly engineer" is in charge of these considerations and package design.

Each of the following stress tests requires 3 assembly lots and 25–30 units per assembly lot. The Package Qualification Test Program needs to be locked down. But, please consult the referenced specifications and/or customer requirements.

4. **Temperature Cycling**
   a. JESD22A-104-C, −55 to + 125 °C
   b. Purpose: Simulate conditions where the devices go through rapid temperature changes stressing the package. Requires pre-conditioning, which simulates part being soldered onto customer's board 3 times. Typical rate of change in the chamber is 2 or 3 cycles of −55 to +125 °C per hour. Read points are generally 350 and 700 cycles.
   c. What do you need?
      i. TC chamber, pre-con capability.
5. **Biased HAST (Highly Accelerated Stress Test)**
   a. JESD22-A110D 130 °C, 85%RH, Duration 96 h
   b. Purpose: Simulate the part operating in high humidity and heat environments.
   c. What do you need?
      i. Biased HAST board where the DUTs are powered up in a "low power state" inside the chamber.
6. **Unbiased HAST**
   a. JESD22A-102D
   b. Purpose: Simulate high humidity and heat environments where the end product maybe deployed.
   c. What do you need?
      i. Oven that can provide 85c and 85% humidity
7. **Bake**
   a. High Temperature Storage, JESD22A-103-D, TA = 150 °C
   b. Purpose: Simulate the End Of the Line bake requirement and expose any marginalities.
   c. What do you need?
      i. Bake Oven
      ii. Reliability Qualification Test Program that has been "locked down".

HELPFUL HINTS:

Some of your customers may ask for additional qualification, characterization or confor-mance to certain standards for a product.

In order to ensure that you do not miss any of these additional requirements you must get from your customer contact a document that provides that customers' requirements for qualification of your specific product.

All major companies that purchase semiconductor based products have documents that describe all the required tests and data gathering to be submitted by vendors before the part is considered qualified by their standards.

Get a hold of that document and read it carefully looking for areas where your qualification plan is missing tests or data that the customer requires.

Create a table showing your qualification plan and additional test needed to close the gap.

You can generally negotiate away some of these requirements depending on the extent of your original plan.

Estimate the resources, cost and time line required to perform these additional tests. Once you have this estimate you can discuss with management the options on how to proceed.

1. Package Qualification Minimum Test Program configuration:

Capability for Unit Serial Number entry into test program so data and unit can be tracked one to one.

Open/Short (Continuity)

Power Supplies Shorts Test

IO Leakage High , Low and Tri-State

JTAG/Boundry SCAN

Basic functional test such as PLL lock.

Basic SCAN test.

Helpful hints:

Check the test program thoroughly with the test engineer to make sure 100% of IO pins and Power are covered and the "correlation unit" values for IO Leakage, continuity etc. make sense (refer to Datasheet).

Serialize and Datalog your correlation units (will not see stress) and stress units and save the STDF files with all measurement values for the tests mentioned. After each test and test point the units will be tested with the same test program to assess the effect of each stress.

Early Life Failure Rate (SLFR) following HTOL is a must as many customers will ask for it and enables calculation of FIT etc. This stress test involved 1000s of units (usually ~ 3000–4000 units) that are stressed using HTOL boards, BI oven and all the same conditions used for HTOL but are only stressed for a short period of time, usually between 4 and 24 h. Following this stress the units are tested using the full production test program. This exercise will provide the PE a more accurate measure of reliability of the product since the sample size is in the 1000s. After all the units are stressed and tested, PE will calculate a FIT rate and report this to the customers.

### 2.2.3 Level 2 Qualification

This test stresses the package solder balls to board interface and solder ball integrity as the component goes though the board assembly process at the customer's manufacturing site. Two methods exist for performing this stress.

Daisy Chain Package Method:

Units are soldered down onto an evaluation board using customer reflow conditions in manufacturing. The boards + units go into a chamber for Temperature Cycling with Condition A. This test goes out to as high as 6000 cycles.

A typical L2 qualification board looks like Fig. 2.3. A PCB (usually 16 layers) with the units being qualified reflowed onto the board using solder wave machine. The board has a header in case there are monitoring signals are being routed there. The common monitoring technique is to daisy chain 8 solder balls through the die (with special "daisy chain" metal layer) and package and then monitor the resistance of this chain. This is done on all 4 corners of the package where the stresses are greatest. The resistance of the daisy chains must not increase by more than 10% during the test which is Temperature Cycle Condition A (TCA).

**Fig. 2.3** Typical 2nd level qualification board layout

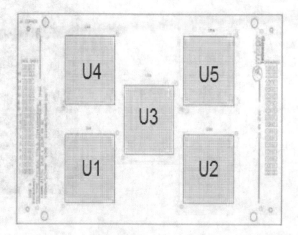

JESD22A-104-C.

Red-outs at 1, 2, 3, 6 k cycles.

As you can see this method will be costly due to development of "daisy chain package" and monitoring circuits.

Dye and Pry Method.

This is a less costly method in that you do not need to develop a "daisy chain package" and there is no monitoring involved during the temperature cycles the boards are going through. The existing product packaged units are reflowed onto a PCB in a certain configuration (probably defined by A customer). The boards go into a TC chamber for certain # of cycles @ certain rate etc. At every read point one board is pulled out. A "certain red dye" is injected between the unit and the board. Following that the dye is cured at a certain temperature and duration. Following this the units are "pried" from the board using a thin chisel and analyzed for any evidence of "solder ball cracks small or large" (Fig. 2.4).

Besides TC there are 2 other tests generally required to complete L2 qualification.

Bend Test:

Bend Test Specification: IPC 9702.

This stress condition basically bends the PCB to a certain degree therefore affecting the interface between ZDD421A and the customer's board. Solder ball integrity is then assessed for a sample of units.

Shock Test Specification: JESD22-B111.

Certain number of boards are put through a drop of a certain height after which solder ball integrity is evaluated for the units on the boards.

**Fig. 2.4** Package Solder balls showing seepage of red dye

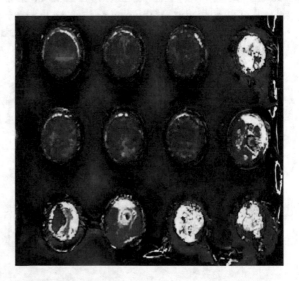

### 2.2.3.1  Planning Qualification Activities
**How to create, review and manage "reliability qualification lots".**
Review the process flow for the wafers (usually 3 fab lots) to ensure it uses the Plan Of Record (POR) process that will eventually go into production.

Once the lot definition is complete send it to your foundry contact so the wafer lots can be started.

- Helpful hint: Once RQ lots are started ask for a provisional fabrication out date.

Product Engineers should get e-test data as soon as possible for these lots. Review the data and chart critical parameters against Control Limits. Ensure everything is in spec. Write a report with charts and summary. Determine which wafers have critical parameters that are close to midpoint between upper and lower specification limits.

Helpful hint: Exclude any "maverick wafers" from qualification lots. These wafers do not represent the plan of record and will likely cause failures that are most likely related to the process excursions on that particular wafer.

Goal: Identify wafers that are as close to typical as possible. Typical $\pm\frac{1}{2}$ sigma is a good target.

Figure 2.5 shows a typical workflow that will produce material required for all the stress tests.

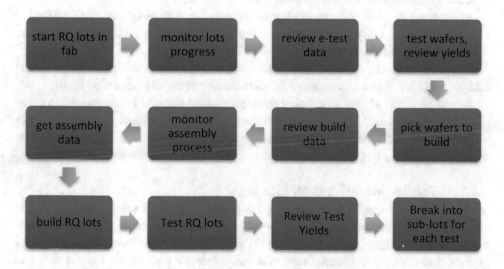

**Fig. 2.5** Reliability qualification lot creation process

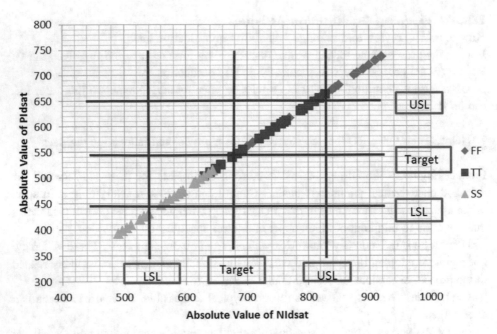

**Fig. 2.6**   Absolute value of N-channel versus P-Channel transistor saturation current example

### 2.2.3.2   **Fabrication Electrical-Test Review Points:**

At a high level all e-test parameters must be compared to their individual test limits using Cpk as a measure. Look for Cpk numbers below 2. Cpk below 2 indicates a parameter that is not in good control or running close to limits.

Plot every critical parameter on a normal distribution curve and place the USL and LSL in the chart so you can compare the distribution vs. the corresponding limits.

Figure 2.6 shows clearly how these critical parameters perform in comparison to upper and lower specification limits of the process for these two transistor parameters. The TT material is mostly within upper and lower specification limits. SS and FF wafers as expected are slower and faster than typical respectively.

### 2.2.3.3   **Fabrication Process Monitoring**

Monitor reliability qualification lots progress during fabrication using a waterfall chart showing the trajectory of the out date. Front of the Line steps (field oxide, well defini-tion, transistor building and isolation stages) progress faster than back of the line steps (metals, isolation, CMP planarization, vias etc.). Ask for post Metal 1 e-test data and look for any parameters that are outside of USL and LSL. Get informed of any wafers that were scrapped during processing and the reason for the scrap. Your fabrication sup-plier must inform you of any misprocessing or difficulties performing a particular step. Wafers with "issues" should be reviewed and possibly excluded from reliability qualifi-cation packaged unit builds. These types of wafers are considered "maverick" because of misprocessing. After reviewing all wafers data the PE is responsible for making the final recommendations for which wafers should be selected.

### 2.2.3.4 Wafer Test Yield Monitoring

Closely monitor wafer test yields for the RQ wafer lots. Ideally you would want the wafer test flow to mimic the final wafer test flow. It is critical to test the wafers at temperatures that the units will be tested at. For example, if final packaged unit test flow is going to be 85 and 0 °C, then test the wafers at the same temperatures. Store the data generated during the wafer test for review and analysis in the network directory structure suggested here.

Compare the yields with what your yield models predict. Highlight any differences between model and actual in your final report for the qualification lots. Differences could be an indication of a couple of different issues.

First suspect is always the test program and the setup used to test the wafers. Make sure you review the setup with the facility used to test the wafers. Sit down with the test engineer and review the test program concerns you might suspect are causing the yield issue. If at the end of this exercise you find and fix any issues with setup or test program, fix those issues quickly, and re-test the failing dice and assess the yield again.

### 2.2.3.5 Assembly Operations Monitoring

Ensure that all steps of assembly operations are performed per supplier specification. No QA or inspection operations should be skipped during the assembly process. Sometimes suppliers skip some or all of these steps to reduce their cycle time. Be specific about compliance to standard process flow. Assembly operations suppliers collect extensive data for the first few lots of a new product going through the line to monitor and eventually qualify their assembly process for ZDD421A. Ask for a summary of that data and any other data collected during assembly of RQ lots. Review this data closely with the package designer and highlight any issues you see and how to approach the ramifications for RQ lots. Deposit this data in it's appropriate location on PE network location.

Monitor assembly yields and assembly data collected by subcontractors to ensure material is close to typical and represents furfure POR.

### 2.2.3.6 Package Test Yield Monitoring

Package test yields should be very close to 98–99% if there are no test program issues. Scrutinize any numbers below these suggested values. If all bad and marginal dice were screened out at wafer test the package yields should be very high.

Helpful hints:

- Review the output of the RQTP ensuring that STDF files with parametric data you requested are being recorded.

- Make sure all the required functional tests are included in the test program.
- Tests to capture key parameter shifts must be included.

Test a few units with the initially released RQTP at planned temperatures and make sure there are no test program or yield issues. If there are any such issues alert TE as soon as possible so root cause and fix can be done quickly. Make sure all the parametric data and data required for shift analysis is collected in proper format for post processing through your YES or other methods to easily create the shift analysis reports. Complete the RQTP checklist shown in Table 2.1 and deposit it in the proper network location. Lockdown RQTP and communicate the completion of this critical stage to all team members. PE can now test the RQ lots at T0.

Stage lots at test waiting for "Reliability Qualification Test Program" (RQTP).

Once RQTP is delivered by test engineer start the test program checkout activities.

- Yields
- Failure pareto chart to resolve any obvious test program issues.
- Parametric Data collection in datalog files enabled and
- Meeting requirements for post processing and charting requirements.
- Ensure all data is retrieved after time zero testing and subsequent time points and stored on the network structure Fig. 2.7.

**Table 2.1**  Die qualification plan Lot#s per stress test

| HTOL | BIASED HAST | HAST | ESD | LATCH-UP |
|---|---|---|---|---|
| QL4200.1 | QL4200.1 | QL4200.1 | QL4200.1 | QL4200.1 |
| QL4200.2 | QL4200.2 | QL4200.2 | QL4200.2 | QL4200.2 |
| QL4200.3 | QL4200.3 | QL4200.3 | QL4200.3 | QL4200.3 |

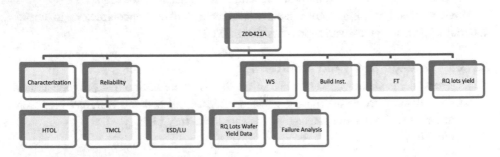

**Fig. 2.7**  PE network data structure

- Checkout the RQTP with a few units from these lots and if no issues run a large sample of units over temperature range $\pm$ guard bands. Release the test program and "lock it down".
- Test qualification lots at data sheet operating temperature extremes + GBs.
- Save and review all the data closely for

Divide passing units from reliability qualification lots into sub lots and units to meet qualification plans in Table 2.2.

Attach RQ test travelers appropriate for each test to its corresponding lots.

Contact RQ test vendors, deliver the units and start running tests in the plan.

Bring back these lots after every subsequent test point and test with the RQTP.

Retain test results, datalogs and STDF files in a repository structured to be organized and easy to access in the future. An example is provided in Figure 2.7

**Qualification Plan Tables**

A typical die qualification plan looks like Table 2.2

A typical package qualification plan looks like Tables 2.3, 2.4 and 2.5.

**Table 2.2**  Die qualification plan unit count per lot

| HTOL | BIASED HAST | HAST | ESD | LATCH-UP |
|------|-------------|------|-----|----------|
| 77 + 5 | 25 + 5 | 25 + 5 | 10 + 2 | 10 + 2 |
| 77 + 5 | 25 + 5 | 25 + 5 | 10 + 2 | 10 + 2 |
| 77 + 5 | 25 + 5 | 25 + 5 | 10 + 2 | 10 + 2 |

**Table 2.3**  Package qualification plan Lot#s per stress test

| TMCL | BAKE | DROP | L2 | L2/SHOCK/BEND |
|------|------|------|----|----------------|
| QL4200.4 | QL4200.4 | QL4200.4 | QL4200.4 | QL4200.4 |
| QL4200.5 | QL4200.5 | QL4200.5 | QL4200.5 | QL4200.5 |
| QL4200.6 | QL4200.6 | QL4200.6 | QL4200.6 | QL4200.5 |

**Table 2.4**  Package qualification plan unit count per lot

| TMCL | BAKE | DROP | L2 | L2/SHOCK/BEND |
|------|------|------|----|----------------|
| 25 + 5 | 25 + 5 | 25 + 5 | 10 + 2 | 50 |
| 25 + 5 | 25 + 5 | 25 + 5 | 10 + 2 | 10 |
| 25 + 5 | 25 + 5 | 25 + 5 | 10 + 2 | 10 |

**Table 2.5**  Typical stress read points for die and package stress tests

| STRESS | READ POINT 1 | RP2 | RP3 | RP4 |
|---|---|---|---|---|
| HTOL | 24 h | 48 | 168 | 500 |
| TMCL | 250 CYCLES | 25 + 5 | 10 + 2 | 10 |
| BAKE | 500 h | 1000 | | |
| BIASED HAST | 96 h | 168 | | |
| HAST | 168 h | 500 | 1000 | |

#### 2.2.3.7  Reliability Qualification Test Program Checklist

The following is a proposed checklist to follow for the reliability qualification test program content (Table 2.6).

### 2.2.4  Assess Your Products' Reliability Using Industry Standard Specifications

In order to assess reliability of a semiconductor based product you first need to understand a bit about what modes of failures most occur in these types of devices.

Let's start from the semiconductor side. There are many "well known failure modes" in semiconductor based products.

Here is a list of what I have learned about and encountered during my career.

### 2.2.5  Electromigration

- This failure mode occurs when current is passed through metal line(s) for extended periods of time (basically the device operating over long periods of time) at densities close to or beyond the limit defined by foundry, causing the metal grains to move or "migrate" to a different location. This phenomenon can result in metal lines becoming open, have higher resistance or in very severe cases short itself to an adjacent metal line (along an existing bur). This type of failure is generally modeled using Black's Model.

$$MTTF = A * J^{-n} * e^{\frac{Ea}{kT}}$$

**Table 2.6** Reliability qualification test program checklist

| | | | | |
|---|---|---|---|---|
| Operating Temperature Range Check with 5 units | | | | |
| Minimum Voltages meet spec. | | | | |
| Minimum Guard bands meet spec. | | | | |
| Operating Temperature Range Check with 5 units | | | | |
| Minimum Voltages meet spec. | | | | |
| Minimum Guard bands meet spec. | | | | |
| Maximum Voltages meet spec. | | | | |
| Maximum Voltage Guard bands meet spec. | | | | |
| Signal Levels for different IO standards meet VIH/VIL | | | | |
| Signal Levels for different IO standards meet VOH/VOL spec. | | | | |
| Propagation Delay times meet datasheet spec. | | | | |
| Setup Times & Hold Times meet Spec. | | | | |
| Circuit speed & performance standards are tested according to datasheet | | | | |
| Key Parameters are being recorded at requested voltage conditions | | | | |
| Ring Oscillators are enabled & speeds recorded at Min & Max Voltages. | | | | |
| memory minimum operating voltage readings are recorded | | | | |
| DSP minimum operating voltage readings are recorded | | | | |
| Critical Logic minimum operating voltage readings are recorded | | | | |
| Maximum Voltages meet spec. | | | | |
| Maximum Voltage Guard bands meet spec. | | | | |
| Signal Levels for different IO standards meet VIH/VIL | | | | |
| Signal Levels for different IO standards meet VOH/VOL spec. | | | | |
| Propagation Delay times meet datasheet spec. | | | | |
| Setup Times & Hold Times meet Spec. | | | | |
| Circuit speed & performance standards are tested according to datasheet | | | | |
| Key Parameters are being recorded at requested voltage conditions | | | | |
| Ring Oscillators are enabled & speeds recorded at Min & Max Voltages. | | | | |
| memory minimum operating voltage readings are recorded | | | | |
| DSP minimum operating voltage readings are recorded | | | | |
| Critical Logic minimum operating voltage readings are recorded | | | | |

A = Area of the metal line(s)

MTTF : Mean Time to Failure

J = Current Density (mA/cm$^2$)

N = current density exponent

Ea = Activation Energy

K = Boltzmann constant (8.617 × 10$^{-5}$ cV/k)

T = Absolute junction temperature in Kelvin

A = constant

Values of Ea, n and A are derived by taking resistance data on a known metal structure while different constant currents and temperature stresses are being applied over a period of time. The resistance of metal line is then plotted and the resulting curve should be fitted into Black's equation and the resulting constants derived. These values are generally available from your foundry manager.

### 2.2.6 Negative Bias Temperature Instability Lifetime Prediction (NBTI)

This is a phenomenon that predominantly affects p-channel transistors in deep submicron. Similar to HCI, carriers are injected through silicon and into the gate oxide where they are trapped. However, in this case these traps can release the carrier when the device is not operating. This characteristic makes this phenomenon very difficult to measure and assess. There is plenty of literature in public domain that goes into details of this. I have one listed here.

http://citeseerx.ist.psu.edu/viewdoc/download?doi=10.1.1.539.8039andrep=rep1an dtype=pdf

### 2.2.7 Hot Carrier Injection Lifetime Prediction Model (HCI)

Hot Carrier Injection is a phenomenon that occurs in deep submicron transistors during operation. Carriers moving from source to drain along the channel created by the gate voltage sometimes are deflected by the lattice and end up being "injected" into the gate oxide or even deeper into the gate material. Once these carriers are trapped inside the gate oxide or the gate they can change the threshold voltage of the transistor or even the work function of the gate material.

### 2.2.8 Gate Oxide Integrity Lifetime Prediction (GOI)

If enough carriers are trapped inside the gate oxide in the right location they can cause a short or low resistance path from gate to silicon. At that point Gate Oxide lifetime is over!

PE will have to estimate the lifetime of each of these types of failures and the shortest lifetime will be the one that determines the lifetime of the product.

The models and parameters for the failure modes should have been determined by the foundry.

HELPFUL HINTS: Ask your foundry manager for these models so you can estimate the product lifetime.

Generally semiconductor product customers require 10 + years of life under normal operating conditions in the field for their end application. There may be exceptions to this for different market segments. So ask your customer representative or marketing manager what the expected life time is for your product.

Assuming a TAF (Total Acceleration Factor) of 100 for 1000 h is equivalent to 12 years.

- 10 years = 87,600 h
- 12 years = 105,120 h

Package Level Qualification tests such as pre-con, temperature cycling, HAST etc. assess how the final product will hold up while going through board assembly and subsequent stresses once it is soldered onto a board and deployed in the field. As discussed in this chapter there are 2 parts to package qualification:

- 1st Level: Bump to Substrate Integrity: Pre-Con, Temp Cycle, HAST, BHAST, Bake
- 2nd Level: Package Solder ball to board integrity.

Ultimately all these tests simulate stresses and operating conditions that will be applied to ZDD421A once it leaves your factory and is deployed in customer systems! These tests are accelerated so that you can get results in 1000 h or less depending on the test.

Pass/Fail results at each point need to be recorded for the units that have completed that read point. Any failures should raise a red flag and MUST be understood, analyzed, localized, root cause determined and fixes put in place to prevent re-occurrence. Activities surrounding this work are some of the most time and resource intensive parts of the reliability qualification process.

Which circles back to what I discussed in this chapter about DFT features that are incorporated into the product during concept and implementations stages. The more features the quicker the path to finding the root cause of failures.

Any tests that show failure(s) must be repeated after the above steps are completed. The repeat material must incorporate the above fixes so it's efficacy can be validated. There are several JEDEC documents that go into much more detail about semiconductor failure modes and how to assess the reliability of products.

## 2.3　Assessing ZDD421A Reliability

As a general rule, due to small sample sizes, there must be zero valid failures related to stresses applied during the reliability qualification execution. One or two "handling" related failures are generally acceptable. One failure in HTOL and 1 failure during package qualification tests is acceptable. Any failures beyond that is a signal that there are possible reliability issues with this version of the design.

Any failures beyond the above number must complete the following verification cycle before it is deemed fixed (Fig. 2.8).

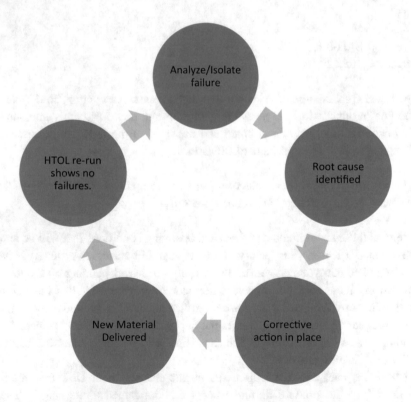

**Fig. 2.8**　Complete failure analysis cycle

## 2.4    Summary

Reliability qualification testing is an essential part of developing and releasing a reliable product into the market place. These series of stress tests will expose any weaknesses in the wafer fabrication, package assembly and testing processes.

Planning and execution of these tests on schedule are critical to the success of any new product. Outside vendors and their services would have to be assessed before they are brought onboard to perform these various tests.

Various stress test, conditions and methods are available to stress both the silicon die, the package and the whole product.

Any failures during these tests have to be validated and moved into failure analysis process. The result of the failure analysis must pinpoint a root cause and enable the design engineering team to put in place a fix to address it. All Test(s) that exposed failures must be redone using material that includes the fix. If the fixes put in place have accomplished their goals there should not be any more failures induced by the stress test.

# Semiconductor Engineering Sample Phase and Product Characterization

**3**

## 3.1 Pre-First Silicon Activities

This phase starts 90 days before 1st silicon arrives. During this period (before all hell breaks loose) an effective PE must prepare and provide the following:

## 3.2 Pre-First Silicon Activities Checklist

1. Request material planners to start Skew lots and backups to those lots.
2. Request material planners to start Reliability Qualification lots and backups to those lots.
3. Provide assembly instructions and top mark for 1st blind build units.
4. BOM and assembly house for this assembly and where is the assembly site.
5. Provide planners the required cycle time for these lots to meet your schedules.
6. 50 units of ZDD421A-M-C 3-Day TAT → Hand carry to HQ
7. 50 units of ZDD421A-K-C 3-Day TAT → Hand Carry to HQ
8. Lot # ZM2561.00 and Wafer # 3 for this build
9. Preliminary Map of the Wafer and identified areas to build from
10. Top Mark Instructions, font selection, logo and print template
11. Example:
12. Line 1: ZDD421A-M-C
13. Line 2: ZM2561.03
14. Line 3: 4220-A2C-ES

---

**Supplementary Information** The online version contains supplementary material available at https://doi.org/10.1007/978-3-031-18030-9_3.

F. Barman, *Semiconductor Product Engineering, Quality and Operations*, Synthesis Lectures on Engineering, Science, and Technology, https://doi.org/10.1007/978-3-031-18030-9_3

15. Line 4: TW
16. Finalize Reliability Qualification plan and schedule.
17. Finalize Manufacturing validation plan and schedule
18. Finalize Characterization Plan and Schedule
19. Create a mask revision list.

Below is a simple depiction of major qualification activities that is easily generated in Excel and can form the basis of tracking the progress of activities (Fig. 3.1).

Post-First Silicon activities checklist.

1. Hand carry Flight 321, FastAir, Depart Taipei 11:40 PM Saturday, Arrive SFO Sunday 4 PM.
2. 1st build unit distributions on WW43D1.
   a. Minimum Test (Continuity, Power Shorts)
   b. By team details
3. Sorted Wafer build distributions on WW45D1.
   a. Package Test Program Equivalent to Wafer Test Program
   b. By team details
4. 1st build for customer samples ready for test WW47D1
   a. Test Start: WW45D1
   b. Test Program name
   c. Test Traveler
5. Create test travelers for every lot at test capturing planned manufacturing flow for ES phase
6. Identify where you will be testing your first 5–10 k units. Reserve tester time at that facility
7. Make sure you have enough package test hardware, sockets (especially spares) and correct trays in quantities needed to meet demand for 3–6 months.
8. Temperature forcing units are also critical for testing at all the planned corners.

## 3.3    First Silicon and Follow on Activities

The first unit builds after 1st silicon has arrived and will require quite some hand holding through the supply chain in order to ensure they are built according to plan of record. Pre-alert all suppliers of what your expectations of quantity and delivery schedule per the plan presented in xxxxxxx.

Creating a mask revision tracking spreadsheet will make it easy to track changes to some or all of the mask layers. Once the chip design has gone through 4 or 5 revisions with different mask layers being affected this document will become invaluable to preserve the history and current revision details.

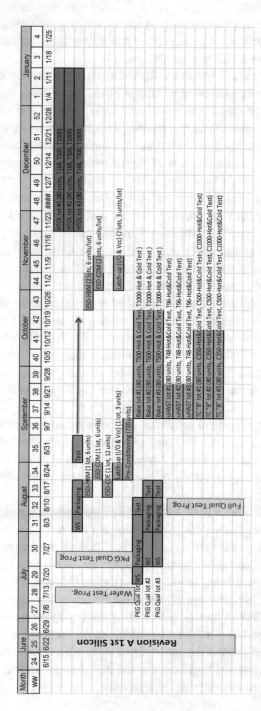

**Fig. 3.1** Simple reliability qualification schedule chart

Finalizing your reliability qualification plan will enable you to provide final inputs to the test engineer so he/she can start developing the reliability qualification test program. This plan will also drive your vendors to provide boards and other hardware required to bring up and execute all the different reliability stress tests on schedule.

In order to have a successful product rollout into high volume manufacturing you will need to finalize your manufacturing validation plans. These plans generally involve starting several wafer lots, assemble the passing dice into packages and putting resulting units through the complete manufacturing process. This includes wafer test, wafer saw, assembly, final test (FT), quality assurance (QA) fine tuning of manufacturing steps and test program optimization. Some portions of the activities may have to be redone to ascertain that any changes made have been effective. Total TAT for these activities can extend up to six months. Therefore extensive planning is critical to success. Sharing the manufacturing validation plan with material planners and test engineers will give them the visibility they need to sync their activities with your plans. Test engineers will need to create the required test programs for wafer and package tests. Material planners will be able to estimate how many wafers they will need to start and how many packages they need to order to meet your required unit volumes for validation.

All these plans and schedules must be reviewed and approved by cross functional teams and their management. Any changes that are found during review must be addressed and plans updated. After the plans are finalized they must be distributed to all groups and be placed on a shared network drive structure defined in this chapter.

I would highly recommend creating a network drive for each product and place related documents there. These locations can be organized using different folders that allow for easy location of documents structure defined in this chapter.

You also must maintain records of reliability, characterization and manufacturing validation data and reports. These are ISO and customer audit requirements. The above structure is ideal for this purpose. Create folders for various activities such as characterization, first customer samples, reliability qualification and internal unit distribution tracking list.

In general there will be a number of lots that will be started once your product "tape out" is completed and mask making has started. Tape out is a term that has stayed in the lingo because in the old days the data required to start making masks was on hard copy tapes or discs! That was a long time ago and nowadays all data is transmitted via secure internet connections to mask making vendors or fabrication facilities. There are not many independent mask making operations in existence any longer as fabrication facilities such as TSMC, Samsung and IBM all have their own mask making operations. One of the reasons for this is that the algorithms involved in making masks for advanced semiconductor processes are specific to the process that will be used in the silicon manufacturing factory. Each mask is biased and prepared based on how the final dimensions should be after the masking and subsequent process steps are performed.

A sample of what are the first few lots to be started is presented below.

1. 1st silicon lot
   a. Usually a small number of wafers so they can be expedited through the fabrication process.
   b. Purpose: Internal engineering samples
2. Backup lot to 1st silicon lot.
   a. In case the original lot is scrapped or does not meet process specification.
   b. Half of this lot will be put "ON HOLD" at the first metal layer in case of any required critical design changes.
3. Customer sample lot.
   a. This is the batch intended to be built for lead customer order fulfillment.
4. Process Skew lot(s).
   a. These are used for characterization, test program development and validation activities.
5. Backup skew lot since risks of not meeting corner targets is high.
6. Reliability Qualification Lots (3 + Backup lots)
7. Additional "risk" material to cover additional customer samples beyond first samples.
8. Manufacturing validation lots. Generally 3 to 5 wafer lots.

Once these lots have been started you must track them in a separate spreadsheet similar to Table 3.1 to ensure that these critical batches are processed promptly and with no major manufacturing issues.

This tracking will also enable the operations team to estimate the number of units available from each batch and whether more batches need to be started to meet projected supply shortfall.

## 3.4 Customer Sample Delivery Planning

For each lot estimate the number of good units that can be produced and how long before good units make it into finished goods inventory. This will enable you to accurately predict the availability and timing of what can be allocated to key customers.

Some customers may request units from skew lots to validate their system. Corner unit requests are tricky to fulfill. Slow units may not pass functional or performance tests. Fast units may exhibit too high of power consumption that is very close to or exceed the proposed limits. Extreme care should be taken to choose or "cherry pick" units so they do not cause issues in customers end system. You may want to consider using acquiring monitor values from candidates and then shipping the units that are within $\pm 1$ sigma away from the average process monitor reading for typical units.

**Table 3.1** ZDD421A supply estimation spreadsheet

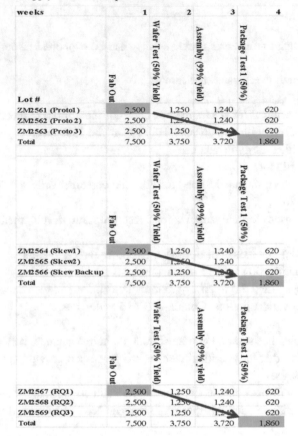

| weeks | 1 | 2 | 3 | 4 |
| --- | --- | --- | --- | --- |
| | Fab Out | Wafer Test (50% Yield) | Assembly (99% yield) | Package Test 1 (50%) |
| **Lot #** | | | | |
| ZM2561 (Proto1) | 2,500 | 1,250 | 1,240 | 620 |
| ZM2562 (Proto2) | 2,500 | 1,250 | 1,240 | 620 |
| ZM2563 (Proto3) | 2,500 | 1,250 | 1,240 | 620 |
| Total | 7,500 | 3,750 | 3,720 | 1,860 |
| ZM2564 (Skew1) | 2,500 | 1,250 | 1,240 | 620 |
| ZM2565 (Skew2) | 2,500 | 1,250 | 1,240 | 620 |
| ZM2566 (Skew Backup) | 2,500 | 1,250 | 1,240 | 620 |
| Total | 7,500 | 3,750 | 3,720 | 1,860 |
| ZM2567 (RQ1) | 2,500 | 1,250 | 1,240 | 620 |
| ZM2568 (RQ2) | 2,500 | 1,250 | 1,240 | 620 |
| ZM2569 (RQ3) | 2,500 | 1,250 | 1,240 | 620 |
| Total | 7,500 | 3,750 | 3,720 | 1,860 |

Keep detailed notes on what test programs were used to screen these candidates and produce the units that are going to be shipped to customers.

Be in regular communication with your foundry manager to get updates on the progress of these lots and make sure they are on or ahead of schedule.

HELPFUL HINT: Create an excel checklist of all the items you need to cover and review it weekly to ensure you are doing what is required to make your product successful and make progress towards your deliverables.

Table 3.2 shows an example.

In ES phase "EXPEDIENCY" is of utmost importance! The whole company has been waiting for this product and now you are the one who needs to make sure that the units are assembled properly in the fastest cycle time possible. Blind-Build + Hand Carry back to HQ is the fastest method but it costs $$$$.

HELPFUL HINTS:

**Table 3.2** Critical actions who, what, when

|  | Who | What | When |
|---|---|---|---|
| Initial test program for blind builds |  |  |  |
| ESD/LU test program |  |  |  |
| Package qualification test program |  |  |  |
| HTOL test program |  |  |  |
| Characterization test program |  |  |  |

Inform management what your cost estimate is to do this.

Alert assembly subcontractor that you will need to do this to save time and get a commitment and quote.

During the early part of the engineering sample phase the PE's first priority should be to accumulate enough good units to be able to distribute to internal teams as well as 1st customer samples. This will be a very challenging phase as the numbers of available good units are limited and there are many teams and customers demanding functional units. First week's priority is to distribute as many units as possible to internal teams so the validation and characterization and other critical activities can proceed without delay. During the 2nd and 3rd week the focus should be to accumulate enough good units to ship to key customers. Customers in general do not expect very high test coverage for this set of units. They just want enough coverage so they can start validating their system built around your chip.

Once you have delivered the above units you will have some breathing room and can turn your attention to preparing for reliability qualification and characterization activities.

Complete your plans for characterization test program and schedule regular meetings with responsible test engineers to align on content and schedule for delivery of the characterization test program (CZTP). Review your plan with him/her and make sure they will deliver a test program that will reliably collect data for all datasheet parameters, functionality and specifications across voltage and temperature.

In the same meeting series you can also discuss the requirements for reliability qualification test programs (RQTP). Delivery date for RQTP is of critical importance since without it no qualification can start. You can also divide this task into two segments. One test program to use after tests that stress the die + package (i.e. the whole product). And a second test program that will be used after stresses that are related to the package. There might also be a place for a very basic test program that can be used after ESD and latch-up tests. This strategy will enable the PE to divide the qualification activities into three smaller segments reducing the impact of a missed deadline. This will also enable the test engineers to build the contents of test program over 2 or 3 stages gradually building the complexity of these critical test programs.

Another topic of discussion for these meetings would be the different versions of the ES test programs that will be used to screen units for customer shipments. In general, major customers require increasing levels of test program coverage over time. Therefore there will be ESV1, ESV2 and so on. Test engineers must agree to deliver these versions on a predetermined time frame. This schedule and deliverables can be aligned to lead customer's development schedule and milestones.

## 3.5    Why Do Characterization?

Characterization is the main tool used by product engineers in order to determine whether ZDD421A will meet product specifications with enough margins over process, voltage and temperature (PVT) during volume production phase.

Skew lots will push the product to a large number of critical process corners to expose and fix weaknesses before the volume production phase. This is of critical importance to the product's future quality, manufacturability and financial success. This exercise will also provide valuable information in centering the fabrication process to optimize yield and quality.

Another reason to complete a comprehensive characterization report is to determine test program sensitivities as voltage and temperature are pushed to extremes. Test program stability, repeatability and reproducibility are key metrics that need to be beefed up during the engineering sample phase. Once these metrics have reached the highest level possible they are ready to be deployed in the volume production phase. Characterization data will also serve as the basis for setting parametric limits in the test program. Characterization and shift analysis data will also determine what gurd bands to use for temperature and voltage. Test program guard bands reduce the risk of inaccuracies in measurement results and test conditions that are collected and set by the test programs respectively.

Customers will also ask for a comprehensive characterization report so they can review the conclusions and feel comfortable that the necessary work has been done to ensure that ZDD421A can reliably run across PVT, meet datasheet specifications with sufficient margins and meet long-term lifetime requirements. The characterization report will demonstrate to them that you can produce a product that meets specifications over a wide range of process fluctuations when in volume manufacturing.

## 3.6    New Product Introduction Pyramid

Very similar to the reliability qualification pyramid, Fig. 3.2 shows the "new product introduction pyramid" that consists of all the activities that must be completed to produce a high quality product. Characterization and reliability qualification create the base of this pyramid. Unlike the reliability qualification, characterization is not a pass/fail type

**Fig. 3.2** New product
introduction pyramid

of activity. It involves collecting large amounts of data over PVT, analyzing that data using simple statistical concepts and exposing areas that need improvement in ZDD421A (Fig. 3.2).

I will cover the creation of "process skew lot" or "process corner lot" in Chap. 4. Units built from these lots are used to characterize ZDD421A. Voltage and temperature variation limits are determined by the product datasheet. However, when setting up your characterization plan, it would be wise to extend voltage and temperature ranges somewhat in order to understand how far the limits of the product can be pushed before it breaks. Temperature extension n should cover the next higher product rating. For example a commercial grade product should be extended to cover industrial temperature range. Voltage extensions are generally in the range of 10–50 mV as there is not a lot of margin built into voltage operating ranges.

Helpful hints: Once your product goes into production, you will always face situations where customer(s) will ask for guarantees that your product will work over extended temperature or voltage beyond what is guaranteed in the product datasheet. If you already have this data you will quickly be able to answer those types of inquiries.

Characterization will also lead to determining where to center the manufacturing process for maximum yield and product performance. If done correctly you can determine whether to skew POR process to one side or other to improve yields or performance.

The basis of your Product Characterization Plan (PCP) is the "Product Datasheet". At times there are also customer specific or product specific requirements that need to be considered. So if you recall the Productization Pyramid below you will see the datasheet and customer specifications form the basis. In the datasheet you should have everything you need to put your PCP together and gauge the margin each and every parameter has with respect to limits in the datasheet.

Every product datasheet must contain the following sections.

1. Product brief or description
2. ZDD421A Block diagram and basic functionality
3. DC Specifications.
4. AC Specifications.
5. AC Timings.
6. Power up and down sequencing.
7. Max/Min Power supply ramp rates.
8. Absolute Maximum Operating conditions.
9. Thermal characteristics.
10. Power supply consumption typical and maximum for all valid operating modes.
11. Voltage regulating and decoupling requirements for power supplies.
12. Input clock(s) requirements for frequency, jitter and input levels.
13. Operating modes, features.
14. Off board flash for FW loading.
15. Register map and tables.
16. Packaged unit Pinout.
17. Package Outline Drawing (POD)
18. Ordering Part Numbers and translation table for all the characters in the OPN.

## 3.7   Basic Concept of Statistical Process Capability (Cpk)

In order to evaluate all the parameters that you are able to measure versus the limits set in the product datasheet we need to use statistical tools such as average, standard deviation and median.

Sample size is also a relevant tool that should be carefully chosen. The more samples you have the higher your confidence in your analysis and resulting conclusions.

Some parameters will need to be modeled or fitted by using linear or logarithmic equations. In this type of modeling regression coefficient (also known as $R^2$).

Cpk is a statistical measure that shows how close a certain parameter is to it's datasheet limit. Basically, you are assessing how far away from $3\sigma$ is the average of your observations from the limit set.

$$\mathrm{Cpk} = \frac{|\mathrm{limit} - \bar{}|}{3*\sigma}$$

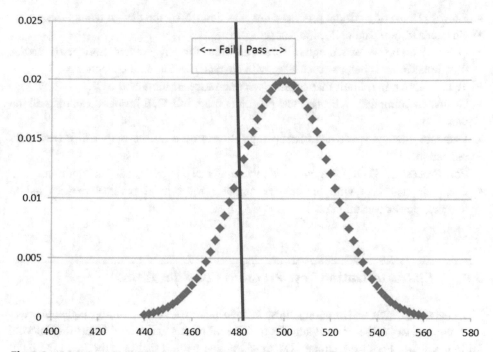

**Fig. 3.3** "Max Speed" distribution versus datasheet limit measured during characterization

As an example let's use an imaginary parameter called "max speed". You have measured this parameter on 10 units × 3 temperatures × 3 voltages = 90 data points. The lower limit for this parameter is 1 GHz let's say. This is a one sided parameter (meaning there is no upper limit). This hypothetical case shows the distribution centered at 500 MHz ($\mu$) and Standard Deviation of 20 MHz ($\sigma$). The lower limit set by the product datasheet is 480 MHz. Anything below this will be considered a "failure" and anything at or above as a "pass" Fig. 3.1.

$$\text{Cpk} = \frac{|\text{limit} - ^-)|}{3 * \text{œ}} = \frac{|480 - 500)|}{3 * 20} = 0.3333$$

Fig. 3.3 shows that once you start manufacturing this product and test 1000 s of units the product yield will suffer by ~15.8% due to certain units not meeting the datasheet "max speed" (Fig. 3.3).

---

## 3.8   Where Does Product Engineer Start All These Activities?

Key requirements to execute a successful PCP (Product Characterization Plan) and produce the required reports, analysis and conclusions are listed below.

- Skew Lot covering critical process corners as defined by the design team.
- Characterization data collection test program
- Close to complete test program to screen units for any defects, parametric and/or functional failures before proceeding with characterization data collection.
- Temperature control unit that covers operating range ± guardband of 5c.
- Characterization data collection test program output in.STDF format containing all the data.
- Database, storage space and backup capability to save all the output files from data collection.
- Post Process of STDF Data collected including analysis by corner and charting.
- Complete datasheet with preliminary numbers for critical parameters such as IO leakage, power consumption.

## 3.9    Characterization Test Program Considerations

The characterization test program must include patterns, levels, timing equations, test flow required to put the product and IOs in different modes and record levels for Vol/Voh, Iol/Ioh, Vil/Vih. For bi-directional IOs, the particular pattern that puts the IO in the Input or Output mode.

This test program will also need capability to search for Vmin for critical blocks such as logic, memory, DSP among many other blocks. Vmin refers to the lowest level that a certain power supply can go before that blocks starts to fail.

Vmin is a critical parameter for many design blocks in advanced technologies. Naturally occurring Voltage variations in systems and boards require that certain amount of guard bands are placed around voltages when the product is tested before it is determined that a unit meets datasheet specifications. Guard bands generally have several components such as measurement accuracy, voltage level application by ATE and finally uncertainty in what is the actual voltage being applied to internal portions of the silicon die. This uncertainty is sometimes referred to as IR drop. Each chip has some certain internal resistance R and consumes a current I so when the chip is in operating mode building blocks may see a lower voltage than what is being applied by the test equipment.. All these guardbands are subtracted from the value specified in the datasheet before it is applied to the power supply. Let's explore an example here.

ZDD421A is specified to operate at a minimum core voltage of 1 v. The power supplies on the ATE have an inaccuracy of 5 mv. The expected IR drop for the whole chip is 10 mv. Therefore the guard band is set at 15 mV. The levels specified in the test program for the core voltage will be 0.85 v. Therefore it is critical to characterize what is the minimum voltage level (Vmin) that the chip can operate at.

Other guard bands are also applied to timing and other levels applied to the chip during ATE testing steps.

Another purpose for collecting characterization data is to determine how to set all the guard bands we discussed here. Once the data is analyzed it is revealed how much margin a parameter has as compared to the limit before it starts to fail. This value will be the upper limit of any guard bands that need to be set. Guard bands for timing parameters are based on the accuracy of the ATE equipment while applying setup or hold times during functional testing.

Based on the operating temperature range specified in the product datasheet a guard band will have to be used to ensure that the values specified are applied to the unit during testing or characterization. Temperature guard bands are mainly based on the accuracy of the equipment applying the condition to the unit during testing. This type of information is readily available from the manufacturer. All temperature forcing units must be calibrated by a certified vendor before being used to test units or collect characterization and other data. After calibration the equipment must receive a sticker with the date the calibration was completed and when it should be calibrated again (usually after 1 year or six months).

Characterization test programs must also include provisions for creating what is called a shmoo plot. This is a type of plot that shows a critical parameter or performance characteristics as a function of voltage or timing variations. This is also known as a contour plot. An example is shown in Fig. 3.4 depicting the passing region shaded in green. This hypothetical case shows that ZDD421A can pass a certain functional pattern as long as voltage and setup time conditions fall in the green region. Outside these conditions that particular functional test will not pass during testing. Therefore the unit will need voltages above 0.65 v and setup times that are between 20 and 27 ns.

**Fig. 3.4**  Shmoo or contour plot of chip functionality/failure versus voltage (Y) and setup time (X)

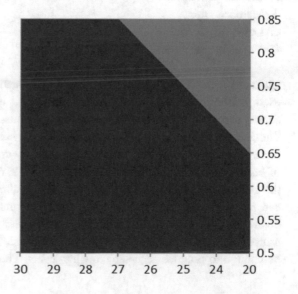

Power consumption for all power supplies including core and IO will need to be collected to ensure that they will meet maximum limits specified in the datasheet. Static (when the chip is in standby mode) and dynamic power must be collected across PVT. ATE environment is ideal for collecting static power data for a product. ATE patterns can be developed and applied that will put the device under test (DUT) in particular standby modes and measure the power consumption using parametric measurement units (PMUs) of the ATE. Measuring dynamic power can also be done on the ATE but this capability is limited due to the particular way a chip is operated in the ATE environment. ATE testing is normally done one block at a time. The whole chip is almost never operating. There are many reasons for this testing strategy which is better left for other chapters. System level test boards are more suitable for measuring dynamic power of the whole chip since the complete chip is always operating in this environment. There needs to be some basic capabilities such as forcing and sensing built into the system level board to enable power measurements for all power supplies.

The output of the characterization test program will also be used to compare key parameters to what circuit simulations predicted. Setup and hold times are a couple of obvious examples. Also of interest might be the max speed for the critical path of the chip. This is of great interest to circuit and block designers since it may expose shortcomings or inaccuracies in their simulations or models. They can use characterization data to adjust simulations or models to better predict product performance.

Test Program content for data collection.

a. IO Char
b. Power (All power supplies individually and total)
c. Critical path speed
d. Memory vmin.
e. Logic SCAN vmin
f. Vmin for any critical analog or digital blocks.

Characterization is not exclusively limited to collecting data on ATE. Another frequently used platform to characterize advanced products is to design a specific board to evaluate certain aspects of the performance such as transmission or receive amplitude and power. These types of boards are particularly helpful to assess performance in a system environment very close to how the product is deployed by customers. PE can also use these types of boards as a second opinion in cases where ATE reports a critical functional failure in the chip. If the validation board reports that the chip is passing running the same test it can possibly point to an issue with the test program or ATE hardware. The validation board can also serve as the basis for a system level test platform during the engineering sample phase. During the engineering sample phase, ATE test programs coverage may not be enough to produce acceptable quality levels. Therefore an additional test insertion in the validation board can augment the ATE test coverage and produce acceptable quality levels.

For advanced technologies (including planar and FINFET transistor based process nodes) it is recommended to insert a stress test at wafer test or final test to expose any latent defects such as gate oxide pin holes, metal whiskers, metal migration or any issues with via and metal connections. Before such a stress is added to production test programs the effect of several stress levels must be studied on units from different process corners to ensure that the stress level is not so high that it is creating additional failure modes. The stress level must be at the right level so only latent defects are exposed. Your fabrication facility representative will be able to provide you with stress level guidelines and methods that are suitable for your specific process node. There are two types of electrical stresses that can be applied to units. Static and dynamic stress. During static stresses, one or more voltage supplies are raised above Vmax for a very short period of time in the order of 100–500 ms and then lowered back to nominal level. Some basic functional or parametric test(s) will be run before and after the stress to collect data on whether the stress has caused a failure in a previously passing test. Dynamic stress tests are similar but the stress is applied while the unit under test is running a functional pattern. A combination of both might be required to maximize the expose of latent failures.

## 3.10  Data Collection

Plan for at least two or three rounds of data collection! I know this may sound excessive but it is a reality. The first round of data collection will generally expose test program issues or shortcomings which must be fixed before data collection can proceed. After required test program changes, the second round of data collection will likely expose data collection or formatting issues that could prevent importing to the yield enhancement system. This would be a show stopper as analysis of this data collected is critical to understanding product shortcomings. After any fixes to address data formatting or other issues the third round of data collection can start. This should be the place you would start collecting data for all the units and conditions until all the data has been collected.

While this data is being collected take a sample of data and put it through the statistical analysis module of your yield enhancement system. if you see any issues at this late stage they should be minor. This step will ensure that the process with the complete dataset will be smooth and on schedule.

The data being collected must be placed in appropriate network locations and backed up regularly to eliminate the risk of damaged or missing data.

If your production data for your products is being housed in a tool such as Quantix (previously known as Galaxy), SiliconDash or Dataview the shortest path to getting your data analyzed is to use the same tool but ask for a "characterization data analysis module".

http://testspectrum.com/home/products/dataview/
https://www.mentor.com/products/silicon-yield/quantix/
https://www.qualtera.com/solutions/silicondash/

Such a module can greatly speed up the data analysis phase for large volumes of data collected. I would venture to say that without such a module PE will not be able to produce a meaningful characterization report.

Place these files in the appropriate directory as described in Chapter Fig. 2.7 data retention structure.

Start the data collection activities by creating lot travelers for all the corner lots so it is clear as to test program, batch number, quantity, temperature and other details are clear and do not have to be repeated due to execution errors. To ensure data integrity for this number of files, I would advise PE to run a few units at two different temperatures and put these files through the analysis tool and ensure there are no issues. Any issues discovered must be quickly resolved so they don't impact the schedule too much.

This activity can take several days and maybe upto two weeks to complete depending on what amount of data is being collected by the CZTP.

Characterization data will be collected at −45c, −25c, 0c, 25c and 125c as this covers the operating conditions stated in the datasheet.

Collecting this data involves running every unit from above matrix with the characterization test program (i.e. CZTP). 5 units per corner, 6 corners run at 5 different temperatures with the same CZTP. This will produce 150 output files in.STDF format.

If all the characterization data is collected in STDF format it can be simply imported into your Yield Enhancement Software and database and analyzed within a short period of time.

An example of how quickly the amount of data collected can explode is outlined below. PVT (Process, Voltage, Temperature).

Every characterization effort involves collecting data over "PVT". That means:

5 units/corner.
12 process corners.
2 voltages.
3 temperatures.
20 parameters.

Total: 7200 datasets to be analyzed!!!! THAT'S HUGE!!!!

The sheer amount of data analysis and charting that must be done dictates that a semiconductor data analysis package with characterization and shift analysis capabilities be deployed.

Several reputable vendors exist to address this need. Features and capabilities greatly influence the deployment, license and maintenance costs. Such a tool will eventually house all the manufacturing and yield data so it will be utilized over the lifetime of

ZDD421A. Careful considerations must be pursued before deciding to purchase from a certain vendor. It is important to have this capability in place well before first silicon arrives.

## 3.11 Data Analysis

Table 3.3 shows a very simple way to quickly assess what parameters from the datasheet are at risk of not meeting the Cpk > 2 guideline.

The methodology described here can be applied to any product parameter that is measurable with reasonable accuracy and has defined limits.

- Create a simple spreadsheet with the data you have collected.
- Calculate Standard Deviation ($\sigma$) of the data points.
- Calculate Mean or average ($\mu$) of the data points.
- Limit for this parameter should be determined previously.
- Use the previously discussed formula to calculate Cpk for all the parameters of interest.
- Any parameter with Cpk value of less than 2.0 must be looked at very carefully to determine what is the reason for below acceptable Cpk????
- In parallel PE must validate the characterization data taken to ensure that all data are valid otherwise the basis of the calculations are wrong and therefore the analysis itself is invalid.

As a result of this analysis several scenarios might result.

- Product Specification is changed so a particular parameter achieves a Cpk of 2 or higher.
  - Generally in the form of "parameter limit relaxation".
- A yield loss of a certain magnitude is incorporated into the costing of the product.
- Certain portions of product design are modified to improve performance of this parameter over PVT. This path will likely lead to a revision B tapeout for certain masking layers.

Every product datasheet parameter, timing and specification must fall within one of the following categories:

- Guaranteed by characterization
  - Parameters that show Cpk greater than 2 over PVT can be called guaranteed by characterization.
- Guaranteed by design

- Design team guarantees that based on simulations, data collected and other methods they can guarantee that these parameters will meet specifications over PVT.

  Helpful hint: Perform a spot check at extreme PVT corners to satisfy yourself and the larger team that the above is true and no further work needs to be performed.
- Guaranteed by test
  - Certain parameters must be tested to datasheet limits during production testing

    to guarantee that all units shipped to customers meet the specification. These are generally performance, timing, power among other specifications.

HELPFUL HINTS:

The data gathering and analysis is only about 40% of the work. The majority of the work is to understand why certain parameters have Cpk of less than 2 and how to address them to achieve Cpk greater than 2. Don't lose sight of this fact! These lower Cpk numbers must be addressed in revision B or they will cause unpredictable yield and supply fluctuations during volume manufacturing.

If your data includes values from all PVT (Process, Voltage and Temperature) corners once you calculate the Cpk from that set of data, you will have a very good idea of where the product weaknesses may be occurring.

Table 3.3 shows a simple set of data for ZDD421A collected over PVT (Table 3.3).

The parameters with Cpk values close to or below 2 in Table 3.3 are red flags.

These parameters must be studied and characterized in order to understand, root cause and fix is put in place for revision B design. If the parameter cannot be fixed, to achieve a Cpk of 2 or better, it must be stated in the final report as a risk to yield, cost and possibly quality.

## 3.12　　Data Presentation and Report Compilation

Follow workflow suggested in Fig. 3.5 to create a clear and concise characterization with valuable conclusions.

Follow Fig. 3.5 workflow to make sure you are covering all the important aspects of a comprehensive characterization report (Fig. 3.5).

The following is an example of how to clearly show characterization data for "max Speed" parameter versus process corners. The red line shows the limit below which the part will fail datasheet specification.

Figure 3.6 was created using a box type chart. Each box shows average, $\pm 25\%$ and $\pm 75\%$

This type of chart combined with datasheet limits is the most effective way to communicate how a certain parameter performs over PVT. Every datasheet parameter must have a chart like above.

**Table 3.3** Typical characterization report summary table

| Parameter name | High limit | Mean | Stdev | CPK |
|---|---|---|---|---|
| Continuity_P | 1.00 | 0.41 | 0.00 | 68.40 |
| Continuity_N | −0.10 | −0.46 | 0.02 | 6.42 |
| Power_Supply_VCCANALOG(AMPERE) | 1.50E-02 | 9.94E-04 | 1.06E-03 | 4.42 |
| LEAKAGE_MIN_JTAG_CLK(AMPERE) | 1.00E-06 | 6.71E-10 | 2.54E-09 | 130.92 |
| DIGITAL_VMIN(VOLT) | 1.00 | 0.76 | 0.00 | 19.65 |
| MEMORY_VMIN(VOLT) | 0.70 | 0.65 | 0.15 | 0.11 |
| ANALOG_PROCESS_MEASURE | 2000.00 | 641.70 | 11.09 | 40.83 |
| TEPERATURE_SENSE_DIODE1 | 120.00 | 109.30 | 2.62 | 1.36 |
| TEPERATURE_SENSE_DIODE2 | 65,535.00 | 3255.30 | 42.55 | 487.86 |
| RING_OSCILLATOR_1 | 2000.00 | 449.60 | 8.43 | 61.29 |
| DSP_PROCESS_MEASURE | 2000.00 | 871.40 | 18.82 | 19.99 |
| DIGITAL_PROCESS_MEASURE | 2000.00 | 832.20 | 13.71 | 28.40 |
| DSNR_1G_Cross | 60.00 | 49.09 | 0.48 | 7.60 |
| DSNR_TXMIT_1G_Echo | −2.50 | −18.15 | 4.85 | 1.08 |
| DSNR_TXMITAmplitude_200MHZ | −0.20 | −1.56 | 0.38 | 1.19 |
| DSNR_ECHO_Amplitude_200 | −0.50 | −0.97 | 0.05 | 3.08 |
| Intersect_Amplitude_150MHZ | −0.40 | −1.73 | 0.14 | 3.14 |
| Intersect_Gain_150MHZ | 64.00 | 60.16 | 0.37 | 3.45 |
| CRITICAL_PATH_A_MAX_FREQUENCY_MHZ | 1000.00 | 400.00 | 80.00 | 2.50 |

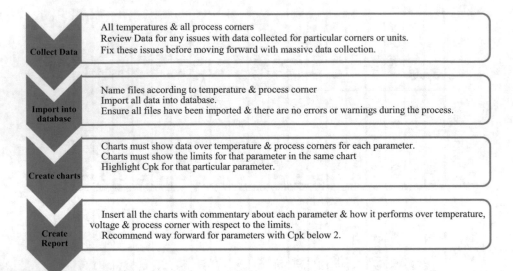

**Fig. 3.5** Characterization report workflow

I can't emphasize enough the importance of creating these charts as they build the basis of how the product will perform and yield over it's lifetime.

What does the above chart tell you about your product's "max speed"?

- Max speed increases as you make the process faster.
- In the slow (SS) corner the product will see some significant yield loss due to max speed specification of 480 MHz (Fig. 3.6).

The final characterization report should also include a summary of skew lot yield and a few example charts to highlight any yield sensitivities observed at process corners.

Here is a simple outline for a characterization report.

(1) One page product brief describing chip functional capabilities and main features.
(2) Define the methodology used for creating the corner lots and what are the details of process variations. For example, your methodology might be that you varied certain transistor parameters and maybe some variations in thickness of metal lines to produce specific process corners. These variations must be decided after discussions and inputs from the design and process teams
(3) One page describing the process variation matrix and which wafers belong t which corner.
(4) Define the test program used to test these corner wafers.
(5) What assembly process was used to build the corner units from passing dice.

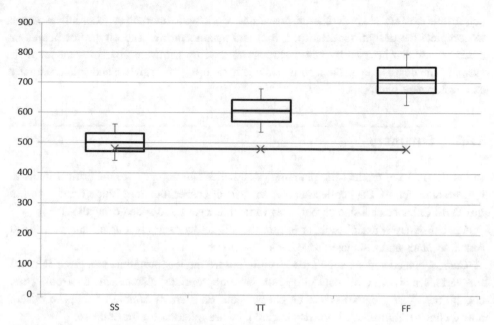

**Fig. 3.6** Critical path speed performance over process corners

(6) What test program was used to screen the corner units and selecting the units that will be used for data collection?
(7) How many units per corner were used for data collection?
(8) What specific temperatures were used to collect data?
(9) What tools were used to post process, store and analyze the data collected from the packaged units?
(10) Include a copy of the datasheet page where the parameters you analyzed are described and what limits apply to each parameter.
(11) Include appropriate charts, histograms and shmoo plots that demonstrate how a particular parameter performs compared to its datasheet limits.
(12) Each chart should be clear as to what parameter is being depicted and what the conclusions of the analysis are.
(13) A summary that gives the reader the key points about how ZDD421 performs in comparison to limits and performance criteria setup in the datasheet.

Other types of characterization involve specifically designed boards to exercise certain features or performance criteria while ZDD421 is functioning in a system environment. These types of boards often include other components such as high speed memory modules, microprocessors and other specialty ICs.

These types of boards can facilitate characterization of ZDD421A with a multitude of other components that customers may use in their systems. A system architect must define this custom characterization board and work with internal or external resources to complete the design, schematics, layout and manufacturing. These types of boards or systems can also be used by key customers to develop firmware and software for their system. The boards can be bundled with some basic bring up firmware and sold separately as development platforms.

## 3.13   Summary

In the ES phase of product life planning of activities and urgency are the critical items to be paid attention to. Data collection and analysis of corner lots must be performed within the Yield Enhancement database system to facilitate ease and speed of analysis.

Checklists are a useful tools to ensure that all activities are resourced and performed based on plans and associated schedules.

Careful considerations should be given to ensure the test programs being put in place for various activities and data collection have appropriate content and data collection capabilities. Test programs must be checked out on a small sample of units to ensure above considerations have been met before they are released for general use.

Using basic statistical concepts will enable product engineers to evaluate key parameters over PVT ranges that meet or exceed datasheet specifications.

# Planning and Delivering Semiconductor Process Skew Lots and Performing Failure Analysis

**4**

## 4.1 Skew Lot Planning and Execution

The basic blocks of each process node are its transistors. Transistors are defined in the front end of the line (FEOL) of process steps. In non-FINFET processes steps like field oxide, well definition, gate oxide, poly gate length etc. define how fast the transistors are going to be and how much power will they consume to deliver that fast performance?

There are many types of transistors available in advanced process nodes. A few examples are HVT, LVT, Native VT, 2.5v IO, 3.3v IO transistors.

Critical parameters for a transistor are Idsat (saturated state drain current), Vth (thershold voltage), Gm (a type of gain), off-state leakage current. In process technologies below 1um minimum feature size the gate oxide thickness had to be reduced to deliver required performance. This led to larger and larger gate current as more advanced technologies were introduced. A certain percentage of electrons or protons goting through the gate oxide get trapped in the oxide and over time incrementally change the transistor threshold voltage.

Each transistor parameter is defined by a typical value (Mean, $\mu$) and standard deviation (i.e. sigma, $\sigma$).

Helpful hint: $+ 1 *sigma$ is approximately $+ 6\%$ away from mean.

Table 4.1 shows a very simple example of how certain key process parameters are moved away from average so they represent slow or fast corners of the process. These types of tables can get very complicated as you add more process parameters to your skew lot definitions.

**Supplementary Information** The online version contains supplementary material available at https://doi.org/10.1007/978-3-031-18030-9_4.

**Table 4.1** Typical process split table for skew lot

|  | Wafer # | 1 | 2 | 3 |
|---|---|---|---|---|
|  | Skew Condition | SS | FF | TT |
| Core NMOS ID sat | STD+6% |  | V |  |
|  | STD |  |  | V |
|  | STD−6% | V |  |  |
| Core PMOS ID sat | STD+6% |  | V |  |
|  | STD |  |  | V |
|  | STD−6% | V |  |  |
| SRAM NMOS VT | STD+10 mV |  | V |  |
|  | STD |  |  | V |
|  | STD−10 mV | V |  |  |
| SRAM PMOS VTH | STD+10 mV | V |  |  |
|  | STD |  |  | V |
|  | STD+10 mV |  | V |  |

Obviously, skew lots are not limited to varying the above parameters. There are many aspects of the fabrication process that can be skewed depending what are key parameters for a certain chip design. Skewing process parameters has to be done in conjunction with process and design engineers since they are the experts on what are the key process parameters and what are their effects on the performance of a certain chip design. The larger the number of variables in a skew exercise the more difficult it will be to determine what is the dominating factor.

An engineer can also skew other parameters such as gate oxide thickness, metal layer thicknesses among others.

Skewing parameters that directly affect transistor performance such as gate oxide thickness, threshold voltage, channel length, doping levels should be considered carefully. Fabrication facilities in general are reluctant to vary these parameters too much from where they have determined to provide optimum performance to a variety of customers. Especially in the volume manufacturing phase since any major changes to the standard process means they have to maintain two different recipes. This will complicate their work to determine whether a process issue or design issue is causing yield loss for your product.

Fabrication facilities can vary the current drive capability of transistors using other methods that do not fundamentally affect the transistor architecture and structure. For example a slight change to the angle of ion implant into the substrate to create source and drain structures can minimally change the effective channel length and therefore transistor performance.

All these adjustments must be discussed with foundry representative and your process engineers to determine the best method to adjust transistor performance.

At times it may be necessary to change metal or isolation thicknesses to improve circuit performance. However, these adjustments will affect the value of additional parasitic resistance, capacitance or inductances.

Before making any changes to transistor or BEOL characteristics many simulations will have to be done by circuit and logic designers to ensure that the adjustments are having the desired effect over voltage, temperature and process corners.

## 4.2  ZDD421A Yield Modeling and Correlation

One of the main reasons for a "process skew lot" is to assess product yield impact at extreme process corners. This method basically simulates future process drifts and their impact on ZDD421A yield and quality. Information extracted from the data collected on these lots will answer the following question:

Are there any changes required in revision B (production mask revision) to address yield variations at process corners?

Are there any changes required to the fabrication process to optimize yield and performance?

Are there any changes to product design and process that must be made to improve quality and reliability of ZDD421A?

Building and analyzing skew batches is very similar to putting ZDD421A through a manufacturing stress test resulting in exposure of areas of weakness. Once these weaknesses are exposed PE will have to present clear and concise data and conclusions that will convince the design team to make modifications to certain mask layers before going into the volume production phase. PEs conclusions must be scrutinized by PE management to ensure it is sound advice and can stand up to challenge. Once that is done it can be presented to a wider audience.

Semiconductor yield models use chip area, defect density (i.e. number of imperfections per square inch of silicon), masking step difficulty level and compression factor and other factors as inputs to predict yield. Compression factor and difficulty level refers to how dense a circuit block is and how that density affects yield. As an example memory blocks are designed with very high density to reduce area. Therefore if the design has large blocks of memories there has to be a higher compression/difficulty factor for the memory blocks when developing the yield model. There are many ways to do yield modeling as you can imagine. Poisson and Murphy's yield models are some of the popular ones. Best place to start is a discussion with your fabrication facilities representative. He or she can provide you with a basic yield model that has been developed by the manufacturer and has been validated by many different products. You can also develop and use your own model based on experience with the previous generations of your particular product.

A reasonably accurate model would take into account areas of major blocks and their percentage of total chip area. Each block should have its own difficulty level assigned to make the mode more accurate. Other factors to take into account are factors such as:

– How tight and compacted is the layout of a certain block?
  • Higher density layout blocks should have a higher difficulty level assigned to them. And opposite for lower density.
– How Many metal layers are used to connect transistors, blocks etc.?
  • More metal layers should have higher defect density assigned to them since higher metal layers involve more chemical and mechanical polishing (also referred to as CMP) stages that will introduce higher level of defects.
– How many process steps are in the whole process?
  • The more process steps the higher the chances of introducing defects and therefore the lower the final yield will be.
– Fig. 4.1 shows the effect of defect density and die size on wafer yield for a hypothetical yield model I have developed for this purpose.

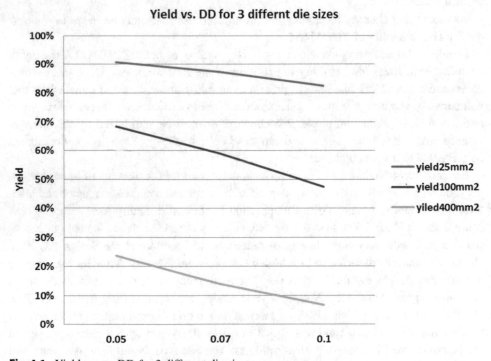

**Fig. 4.1**  Yield versus DD for 3 different die sizes

As you can see defect density has a dramatic effect on yield especially for larger die sizes. Therefore it is advisable to put pressure on your wafer supplier to put together and present a plan and schedule for defect density reduction especially in the volume production phase. The plan must include specific actions to address certain defects that the supplier has observed in that particular process node. Once the plan is in place make sure that you review it with the supplier on a quarterly basis and track the DD improvements plan and how it is improving the overall yield of ZDD421A product.

Reviewing a product's yield trend over time is essential in determining whether that product is achieving the defect density and yield expectation provided by the wafer fabrication vendor and the yield model developed by the product engineer. Below are some of the reasons this maybe happening.

(1) The defect density assumed in the model is not accurate and the process has higher levels of defect density.
(2) The model does not take into account one or more factors such as difficulty level of a certain design block.
(3) The product may have deficiencies or sensitivities in one or more areas resulting in lower than expected yield.

To illustrate the point about yield model comparison to actual yields I have put together Tables 4.2 and 4.3 the yield model equations behind this data is not that important to delve into in any detail. The important points are where the actual yield deviates significantly from the yield model raising a RED FLAG.

Why is there such a large difference? Is it a product test or process issue?

What approach should be taken to understand, root cause therefore reducing the yield variance?

Depending on the type of failure(s) causing yield variance the failure analysis approach will have to be changed. Parametric, performance and functional failures will each require a different approach. I will discuss the different approaches later in this chapter.

ZDD421A is 10 m× 10 mm repeated on a 12 inch wafer consisting of 576 total sites. Difficulty factor of 5 and compression factor of 10 is assumed. Digital area is 30% of total chip area. Memory is 35%. DSP is 25% and analog is 10%. This model estimates that digital area fallout should be no more than 21%. Memory fallout should be no more than 13%. DSP area should have no more than 9% fallout. Analog area should have no more than 4% fallout. Fallout is determined by whether the failing chip has failure(s) in digital, memory, DSP or an Analog test during wafer test.

Figures 4.4 and 4.5 show a clear chart summarizing parametric or functional fallout of ZDD421A by process corner in comparison to the yield model.

This analysis reveals that the TT process wafer shows higher than expected fallout for all blocks. SS wafer has even worse results compared to TT and model. Digital block shows it has a weakness if the process drifts towards the slow corner. FF and FS wafers

**Table 4.2**  ZDD421A yield model versus actual skew lot fallout and yield for TT and SS wafers

| | Yield Model | Model Expected Fallout% | First Silicon TT Fallout% | SS Wafer |
|---|---|---|---|---|
| Die Width (mm) | 10 | | | |
| Die Length (mm) | 10 | | | |
| Chip Area (mm2) | 100 | | | |
| Chip Area (sq inch) | 0.155 | | | |
| Reff | 0.25 | | | |
| Wafer Diameter (inch) | 12 | | | |
| Gross Die Per Wafer | 576 | | | |
| D0(Defects/Sq. Inch) | 0.05 | | | |
| Dificulty factor | 5 | | | |
| Comp. Factor | 10 | | | |
| Digital Area (%) | 30% | | | |
| Digital area D0 factor | 2 | | | |
| Digital area Yield | 79% | 21% | 30% | 50% |
| Memory Area (%) | 35% | | | |
| Memory area D0 factor | 1 | | | |
| Memory area Yield | 87% | 13% | 12% | 24% |
| DSP Area (%) | 25% | | | |
| DSP area D0 factor | 1 | | | |
| DSP area Yield | 91% | 9% | 13% | 25% |
| Analog Area (%) | 10% | | | |
| Analog area D0 factor | 1 | | | |
| Analog area Yield | 96% | 4% | 10% | 20% |
| Overall Yield | 61% | | 35% | 23% |
| NDPW | 350 | | | |
| Block Area as % of total area | | | | |
| Digital | 30% | | | |
| Memory | 35% | | | |
| DSP | 25% | | | |
| Analog | 10% | | | |

show results that are close to what the yield model has predicted. SF wafer is in line with the model with the exception of memory having larger than expected fallout. The memory block has a weakness if n-channel drifts towards the slow corner.

Overall ZDD421A has yield issues at SS and SF corners. FF corner yields close to yield model but there may be other yield issues in this corner once wafer is packaged into finished units and ambient temperature raised to 110c in order to reach junction temperature of 125c specified in the datasheet.

The first order of data analysis is to have a chart that shows a particular parameter at different process corners (SS, TT & FF for example).

Also of interest would be to show the yield fallout per key block at all process corners. It would look similar to Fig. 4.2. This will make it clear to the audience how yield varies as process drifts from one corner to another. Looking at Fig. 4.2 it is quickly clear that memory functional test has higher than expected fallout at SF corner. Also you can quickly see that scan_TDF_Vmin has higher than expected fallout at SS corner (Fig. 4.3).

**Table 4.3** ZDD421A yield model versus actual skew lot fallout and yield for FF, SF and FS wafer

| | Yield Model | Model Expected Fallout% | FF Wafer | SFWafer | FS Wafer |
|---|---|---|---|---|---|
| Die Width (mm) | 10 | | | | |
| Die Length (mm) | 10 | | | | |
| Chip Area (mm2) | 100 | | | | |
| Chip Area (sq inch) | 0.155 | | | | |
| Reff | 0.25 | | | | |
| Wafer Diameter (inch) | 12 | | | | |
| Gross Die Per Wafer | 576 | | | | |
| D0(Defects/Sq. Inch) | 0.05 | | | | |
| Dificulty factor | 5 | | | | |
| Comp. Factor | 10 | | | | |
| Digital Area (%) | 30% | | | | |
| Digital area D0 factor | 2 | | | | |
| Digital area Yield | 79% | 21% | 22% | 23% | 23% |
| Memory Area (%) | 35% | | | | |
| Memory area D0 factor | 1 | | | | |
| Memory area Yield | 87% | 13% | 10% | 50% | 10% |
| DSP Area (%) | 25% | | | | |
| DSP area D0 factor | 1 | | | | |
| DSP area Yield | 91% | 9% | 8% | 10% | 12% |
| Analog Area (%) | 10% | | | | |
| Analog area D0 factor | 1 | | | | |
| Analog area Yield | 96% | 4% | 5% | 20% | 12% |
| Overall Yield | 61% | | 61% | 28% | 54% |
| NDPW | 350 | | | | |
| | | | | | |
| Block Area as % of total area | | | | | |
| Digital | | 30% | | | |
| Memory | | 35% | | | |
| DSP | | 25% | | | |
| Analog | | 10% | | | |

Since my simple model docs not take into account variations of voltage or temperature the rate of failures at different temperatures and voltage conditions must be looked at carefully by testing the same set of wafers at temperature extremes and assessing the yield variations. Some of the questions to ask after that data has been collected is:

How are these failures affected at Vmax and Vmin?

How are these failures affected by temperature when varied from −45c to + 125c?

How are these failures affected by the process corner? Do slower dies have more failures?

When analyzing such failures it is very important to assess where the particular die stands in the spectrum of process corners. That's where having the electrical test (e-test) data for that particular die is of utmost importance.

This can be achieved in a couple of different ways. Let's look at a couple of options.

Option 1:

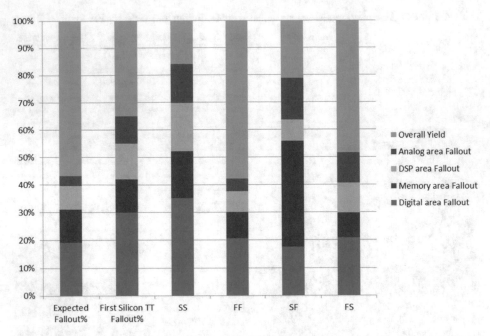

**Fig. 4.2** Key product block yield model fallout versus actual skew lot fallout rate for TT, SS, FF, SF and FS process corners

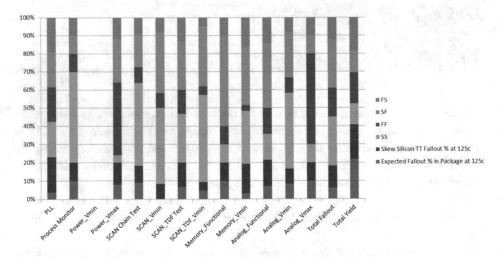

**Fig. 4.3** Key parameter and key test failures versus process corner chart

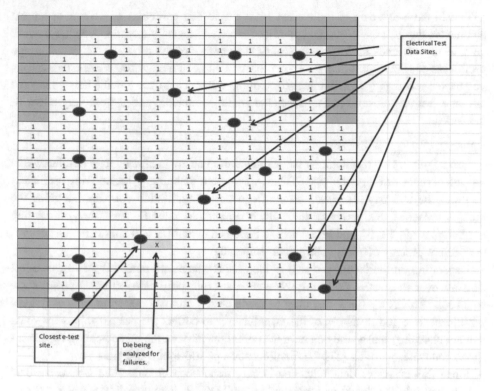

**Fig. 4.4**  ZDD421A wafer map and location of electrical test sites

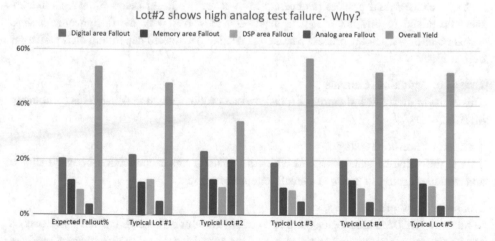

**Fig. 4.5**  ZDD421A lot to lot yield variation chart

Obtain the e-test data of the closest e-test site to that particular unit. Look at critical parameters like Idsat, Vth etc. measured on that e-test site and determine whether the unit being analyzed lands in the fast, typical or slow corner of the process. This is where the usefulness of 100% e-test data for corner lots will come into play. 100% e-test data means you have e-test data for the whole surface of the wafer. Therefore any die being analyzed has an e-test site close to it. Pay special attention to including sites from very edges of the wafer.

Option 2:

Use a built in ring oscillator on the particular die you are analyzing to determine which process corner this unit belongs to. This is where your planning to include ring oscillators on every die will pay off. Based on the frequency of oscillation you can determine the process corner of this particular die. Circuit simulations must be performed on the ring oscillators using TT, SS and FF models so there is a model for how these circuits perform at different process corners, voltage and temperatures.

Depending on the layout and size of the product, there are potentially 100 s of e-test sites all over the wafer. Collect 100% e-test data (not only selected sites) on all skew wafer lots. This is a request that must be made clearly and well ahead of time as it takes significant time and resources of the manufacturer to collect the data. If this is the route you are choosing, make sure you have the commitment from the manufacturer. Figure 4.4 shows the wafer map for ZDD421A and location of 100% of e-test sites.

Since this extensive data collection is only done once or twice, make sure the data is imported into the "yield management system" (YMS) including e-test, wafer test and package test) for analysis and correlation between all three test stages.

Key transistor and process parameters to look for are listed below. Other parameters must be added to this list based on detailed discussions with the design and process teams regarding what key process parameters need to be tracked depending on a particular design block.

Transistor Saturation Currents

This is the highest level of current a transistor can produce when the channel is completely inverted.

Transistor Threshold voltages

This is the value of gate voltage at which a transistor has a connection between source and drain and starts to produce a small amount of current.

Transistor Subthreshold leakages

In advance CMOS technologies there is a certain amount of leakage current that exists as soon as a voltage is applied to the drain of the transistor. Even though there is no gate voltage the junction between drain and substrate will start leaking current.

Spacer widths
Advanced CMOS technologies deploy deposited spacers to control the doping in the source and drain of transistors to reduce parasitic resistances and improve the performance.

Gate Oxide thicknesses
The gate oxide is what isolates the conductive gate material from the conductive substrate allowing the gate voltage to control the conversion of the channel from one type to another therefore connecting the source and drain and allow current to flow from source to drain of the transistor.

Contact resistances
Contacts are what allow a metal layer to connect to other structures in the chip such as transistors, capacitors or other elements of the design. Higher via resistance will hamper the flow of current therefore lower overall performance of the chip.

Via Resistances
A via is similar to a contact but it is used to connect conductive layers that are on different planes. For example if you want to connect metal 1 to metal 3 you put a via that starts on metal 1, is isolated from metal 2, and makes a connection to metal 3. These are crucial in laying out circuits that are compact. As you can see higher via resistance will reduce the current flow and therefore decrease performance of the block or a complete chip (Fig. 4.4).

Figure 4.5 shows a case where typical lot #2 yield loss is higher for analog tests as compared to other typical lots. This should raise concern for product engineers and warrants a closer look. Analog tests are very sensitive to quality of contact between package solder balls and load board socket. Any unexpected additional resistance can adversely affect sensitive measurements like voltage and current. Most of the time these types of failures are traced to dirty socket pogo pins or dirty solder balls on the package. Cleaning of pogo pins and solder balls with appropriate brushes most likely will fix the issue.

## 4.3    Edge Effects and Yield Pitfalls

The skew lot captures all these possible variations in one (or sometimes 2 lots, depending on # of splits). These lot(s) and resulting units will enable all product team members to characterize their block and its performance over the complete process variation space. The key here is to see what parameters or performance measures will "fall off the cliff" at which process corners.

After transistors are defined the process moves into the Back End of the Line (BEOL) where transistors are connected to create circuits, blocks, inter-block connections, IO implementation and other portions of ZDD421A. This is achieved by building many layers of metal, contact, isolation, via etc. that intricately connects everything together to create the full chip. It is not uncommon to have 10 layers of metal in an advanced process node.

Depending on the specific design PE might decide to expand the set of characteristics to include characteristics that are appropriate for analog, memory or mixed signal circuits.

Semiconductor processing relies heavily on the polishing process in planarizing materials deposited using CVD so the top surface is flat before another layer is created on top of it. It is very similar to surface grinding in the metal industry to make sure surfaces are flat and void of any imperfections. Polishing generally involved orbital movements by the tool to flatten all surfaces. This means that there are phenomena that are affected by this orbital movement. A very common effect seen when analyzing a single wafer yield is called the doughnut effect.

Each subsequent circle after this could show poor yield due to certain failure modes. The extreme outside circles of die are very susceptible to what are called edge effects. Pay special attention to how the e-test parameters change from center to the edge. Lithography, polishing, deposition and other process steps lose their consistency when they process edge portions of each wafer. Depending on what process step is affected transistor saturation currents could drift higher or lower at the edges.

A simple way to assess edge effects is to look at how far away from wafer average is a particular e-test data or ring oscillator performance. Is it 1, 2 or 3 sigma away from average? The higher the number the more edge affects you are observing.

## 4.4    Automotive Product Considerations

Automotive level products require a much higher level of quality (sometimes called zero defect) and process control because if an automobile IC component fails, in extreme cases, it might lead to stranded customers or worse by causing accidents. Therefore automotive products sometimes require passing dice that are in close proximity of a failure must also be considered failures and not be assembled into packaged units. Another consideration for automotive products is to how thick of a band adjacent to the outer edge of the wafer is excluded from being assembled into packaged units. A commercial grade product may require 2–3 mm exclusion zone where an automotive grade product will require 3–5 mm exclusion zone.

Yield enhancements and analysis software are critical to have in order to enable product engineers to be able to quickly and efficiently do this type of analysis and mapping capabilities for automotive products. All these restrictions will lower automotive product yields and eventually increase the cost of manufacturing such products.

Almost all automotive grade products will require tri-temperature testing until it's proven that cold temperature testing can be eliminated. Automotive customers may require that any test step elimination not add any additional DPPM to product quality levels.

### 4.4.1 Data Presentation Lessons Learned

A simple chart to illustrate how the actual skew lot compares to the plan is shown in Fig. 4.6. This chart shows the N-channel and P-Channel saturation currents with respect to where their respective upper and lower specification limits are. These types of charts are well suited for situations where multiple process corners are to be displayed in one chart.

I have also found histograms to be particularly useful for showing critical parameters that have tight distribution around averages versus parameters that show wide distributions.

HELPFUL HINT: Analyzing wafer test failures is simplified tremendously by packaging a sample of failing die so they can be more easily handled and analyzed in a more controlled temperature environment.

A simple conclusion after analyzing the above data might be summarized as below:

Move n-channel and p-channel transistors towards the fast corner of the process by 1/2 to 1 sigma to reduce chances of any yield excursions in the future.

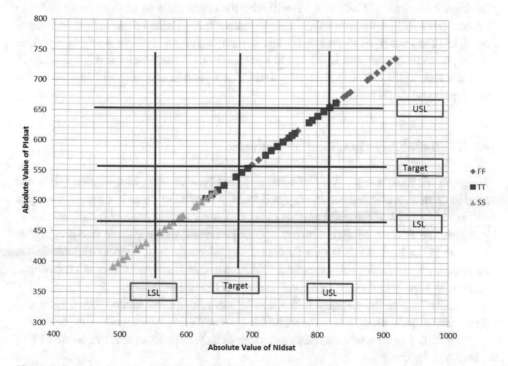

**Fig. 4.6** Sample chart of actual ID sat versus targets for skew lot

Caution with this proposal is that moving the process towards the fast corner will adversely affect ZDD421A power consumption. This will have to be looked at more deeply in packaged units tested to datasheet temperature range.

Present a problem statement for each weakness discovered. Then summarize and present the data collected to clearly illustrate the issue discovered. The analysis must be thorough and the data presented relevant to the issue. Data collected must be scrutinized for any potential issues in the raw data or anything related to the test program used to collect the data. There should not be any questions about the validity of the data, collection method, analysis or conclusions reached. Put forward well thought out proposals for addressing the weaknesses exposed by skew lots.

When you see a failure rate that goes above estimate it MUST RAISE A RED FLAG for you to start looking at why???? Is it a product design issue, is it a process issue, is it a test program issue?

### 4.4.2  Big Picture Goal of Failure Analysis

During the lifetime of ZDD421A PE will have the opportunity to analyze many different failures from different stages of the manufacturing flow. These failures must be categorized and then analyzed to root cause and corrective actions. Failure Analysis activities ending in corrective actions is the key to a robust product (Reliability and Yield related).

Let's look at the different types of failures that a product engineer will receive during the lifetime of ZDD421A.

### 4.4.3  Reliability Qualification Failures

These are the most serious types of failures during New Product Introduction (NPI) and need to be treated with utmost urgency and force to resolution. Here is a generic flow that will work for most of the cases that you may encounter during product qualification.

Verify the failures on ATE yourself and make certain that the same test program as in T0 is being used. Serialize all units and log all results and keep as records in the suggested product folder structure under "Reliability-Qualification" folder as an example. Make sure your "standard or reference" units pass the test program.

Visually inspect the package and specially the solder balls and/or leads. Note any discoloration or damage.

In an ESD safe environment, clean the solder balls with acetone and a horse hair brush and re-test the unit.

Do a complete (all sides and edges) visual inspection with a $10 \times$ or more capable microscope. Observe, note and take pictures of any physical damage(s) or changes.

If no mechanical issues are observed proceed to electrical analysis of the failures.

Which block is exhibiting failure?

Is the failure solid or changes behavior with changes in temperature and/or voltage?

Are there any dependencies on test pattern timings or voltage levels set?

Test the units 5 times to make sure test results are consistent.

Enable "continue on fail" and collect results of every test in the test program. This will tell you a lot more information about the type of failure and how to proceed.

Were there any unusual events during the reliability test?

Was the qualification test conditions specified in the request followed exactly?

Inform all team members of failures once they are verified to be valid and you have gathered enough information to communicate an intelligent assessment of the failure(s).

Inform your suppliers of the failures and share whatever is appropriate in your judgment with them. You will need to put lots of pressure on your suppliers to react to these failures and dedicate resources to the resolution process. Time is of essence here!!!!!

Assess whether this may be a "package related", "silicon related" or "other material related" failure?

Work very closely with your DFT and design team to isolate the failure quickly. This is crucial in ultimate root cause analysis and corrective actions.

This is going to be a pretty painful process because you have to get down to a few transistors (out of millions!!!!! Think about that!!!) before you can start physical failure analysis.

### 4.4.4 Failures During Engineering Validation and Other Internal Group Evaluations

Once units are given out to internal engineering teams they will be deployed in a variety of environments and will see varying levels of mechanical and/or electrical stresses. Be careful how you approach these units because they could have been put under extreme stresses!!! But you should also not ignore these types of failures because they may be the indication of a marginality or manufacturing issue. This is where your engineering judgment and product knowledge comes into play!!

### 4.4.5 Wafer Sort, Final Test or EQA Failures

The first two categories (Wafer sort and final test) are considered yield related failures. The last category, (Electrical QA) is considered a "quality related" failure and is very serious. This could mean two things, both of which are not good. In one scenario you are allowing "marginal units" to get through the final test stage. Meaning you are very likely shipping "marginal units" to your customers!!!! Yikes!!! In the second scenario your test process may be damaging the unit!! That's not a good situation at all!!! In short these are serious failures. First priority is to ensure units are not being damaged during test. PE can use a method called "hammer test" to rule out unit damage during test. In this type of test the unit under test is tested repeatedly (anywhere from $10\times$ to $50\times$) to see if there is any resulting degradation or failure. Tests that draw large amount of current through one or more power supplies are the first place to look for root cause of such damage.

Wafer sort failures could be related to prober setup, probe-card, test program or product itself. Always start with prober/probe-card and work your way down the list. Ultimately, if you determine, these are valid failures they will have to be built into units so they can be easily and quickly analyzed. Once you receive these units, run them through the product test program and gather all the data regarding failing test(s) and other information necessary to localize the failures. Focus on no more than 3 units to speed up the work. Always compare results with "standard units". Do temperature, voltage, speed and time analysis on the failures to separate "solid failures" from "marginal failures". Wafer sort failures must be assessed against your previously developed "block level yield model" to assess where the model and reality diverge. These are generally the weak points of the product. Determine as best as you can which process corner these units are from? SS, FF, FS etc.

Final test failures could also be related to setup, hardware, test program or product related. Basically the same process as wafer sort failures can be followed to root cause and corrective actions. Generally, final test yields are expected to be in the range of 99% or higher. If your product is not in the ballpark of this number you must look into why!!!

Highest level of priority should be given to failures that occur during the manufacturing validation phase. Start with the failure with the highest failure rate and work your way down the list. EQA failures are the highest priority followed by final test and wafer test failures. Work closely with test engineers to analyze verified failures and come up with a plan and actions to resolve these issues quickly. This will improve yield, lower cost and increase the quality of your product dramatically.

## 4.4.6  Customer Manufacturing Site Failures Isolated to Your Product

This is code ORANGE!! This means the units passed all the manufacturing test steps, shipped to the customer, were built into systems and failed the customer's manufacturing test(s). Jump on these ASAP as they could be an indication of quality issues!! I can't stress this any more than I have. Get as much information from the customer about the failure mode and symptoms. As soon as prudent, ask the customer to send the unit and preferably the full system that failed to you for further analysis (basically issue a Returned Material Authorization—RMA). Once the unit is received either tests the individual unit on ATE or brings up the full system along with the engineering team and start the analysis. In the ES phase test programs are not mature and therefore lower the quality level of units going out to customers.

Ensure that these types of failures are tested with the original version of test program that was used to test this particular lot as well as the latest more complete version of the test program. If the unit fails with the latest version of test program you have a very good explanation for this failure.

### 4.4.7   Failures from the "Field" Isolated to Your Product

This is code RED!!! The unit was built into the customer's system, passed the customers' system test, shipped to their end customer and then after sometime failed "while deployed in the field". This is a HUGE problem!! Jump on it like it's a ticking time bomb!!!!!

Start by looking at 1 or 2 verified field failures to localize the failure to a small block and then bring in that block expert and ask for their advice on what to further isolate the root cause and what to do next to localize a few transistors(s). Further analysis may involve creating special test patterns that could provide more information about the failure and it's root cause. It is important to involve the design and test team early so they can provide their inputs and work with you to isolate the root cause and put in place required corrective actions.

In parallel, start looking at any electrical, thermal or environmental stresses that may have been put on the parts during board manufacturing at your customer's site. This is an unlikely source but amy expose areas where undue stress are put on the units during board assembly or testing.

### 4.4.8   Analysis Techniques, Methods and Considerations

Visual Inspection under $10 \times$ or better microscope.

This technique is used to find gross issues on a package level such as solder ball damage, package substrate damage, top side damage, marking accuracy, marking legibility.

X-ray and CSAM

These techniques enable a deeper look into the package and die to package interface using either x-ray or Confocal Scanning Acoustic Microscopy. These methods are useful for cases such as package delamination, die chipping or crack, alignment of die to package.

http://www.muanalysis.com/techniques/confocal-scanning-acoustic-microscopy-csam

### 4.4.9  Basic Power Up Emissions Test

This technique is very useful in getting an early indication of "hot spots" on a failing chip. By applying a small amount of voltage on a particular supply and observing the emissions from the failing unit detected by a sensitive lens on the equipment. This technique is good for situations where large leakage current(s) exist due to damage or defect. Transistor level defects and or leakage cannot be easily detected with this method due to low level of operating current in transistors. Generally this method can detect failing units where the defect or damage has caused a short between 2 isolated points.

### 4.4.10  FIB (Focused Ion Beam)

This is one of the most expensive failure analysis techniques available due to the cost of the equipment and the skill level of the technician who runs this equipment. In this method you will need to narrow down the failing area to a few transistors. Afterwards you can use FIB to perform micro-surgery on the area to cut a metal line and/or make a connection that does not exist. Using this technique you can also open a window in silicon going down a few layers to inspect a suspected damaged or failing area.

### 4.4.11  Layer-By-Layer Removal and Inspection

This technique utilizes grinding wheels or variable speed polishers (and sandpaper) to remove material and in the process providing you a clear and detailed view of a certain area covered by other layers. This technique requires some process knowledge of material composition and thicknesses. Due to time consuming and high accuracy nature of this method it can be costly. The technician's skill level is also a determining factor in the success of the job.

General preparations or data needed for a successful failure analysis job.

It is always good to prepare for any failure analysis job before you actually submit your samples to a lab for failure analysis. This will ensure you don't spend precious time or money without getting any results.

You will need to prepare a layout database of your product that contains a few top metal layers and polysilicon (poly) so the FA lab can use it as reference when looking at your sample under the SEM or FIB machine.

Always provide the lab a few dummy samples of the same type of unit that they can use to experiment on before doing any work on the actual failing sample.

Here are a couple of examples of unusual failure analysis cases that illustrate how to approach different cases with open mind.

## 4.4.12  Real Life Examples

A number of years ago I was working on a brand new product that from day one had much higher than expected power consumption with some lots showing values that were 2× or 3× higher than expected. The product worked as intended and was passing a significant number of validation and functional verification steps. However, this higher than expected power consumption, was preventing us from starting HTOL stress because the boards for this test were not designed to handle 2× or 3× of expected power consumption. Initial failure analysis involved looking at the final test hardware, test patterns, voltage levels and other usual suspects. This particular product was fairly large (close to 50 mm × 50 mm die) and utilized 11 layers of metal to achieve power distribution goals for a large die. Due to so many metal layers stacked on top of one another it was nearly impossible to do any back side emissions and detect a location for where this excessive current. We decided to move onto to destructive analysis to root cause the problem. We did not locate any issues during the first section of layer by layer removal and observation. At one of the metal layers we found a minimum space metal layer, used for power distribution, very tiny metal burrs that were extending from power to Vss lines causing basically a very high resistance short between power and ground. These shorted lines were not enough to affect functionality but were basically creating paths for unnecessary current to be introduced during functional testing. These metal layers had to be changed to increase the distance between power and ground. Also the masking and etch steps that defined the metal lines and spaces had to be optimized so these burrs were eliminated.

Another case involved certain types of IOs not meeting leakage specifications. At first we suspected issue with the IO design or issues with transistors used in those types of IOs. During the initial debug process which always involves test hardware, test program and functional patterns it turned out that these IOs required a certain PLL to be locked and providing a stable output clock before they were able to be driven to high or low levels. That particular PLL required a jumper inserted on the DUT board. This jumper was missing causing the PLL to not lock.

In another case we were looking at some time very large failures in one of the DSP block in units assembled from particular fab lots. The strange thing was that these die had passed the exact same test at wafer level one or two weeks ago! Therefore we suspected something in the assembly process causing these catastrophic failures. From the electrical failure analysis of a few packaged units we localized the physical failing location to DSP blocks located along the very outer edge of the die adjacent to the scribe lines. During destructive failure analysis and inspection of areas near the failing cells we found out that there is mechanical damage that extends from the scribe line into active circuitry just beyond the scribe line boundaries. This made it very likely that during wafer saw operations (after wafer test and before assembly starts) damage was being introduce into the scribe line and subsequent steps propagated that damage into active circuitry causing electrical failures. We worked with our assembly subcontractors to optimize the wafer saw operation to the point where any minute physical damages were contained within the scribe line. These process optimizations significantly reduced the incidents of this type of failure to a tolerable level.

In general PE must to collect as much information as possible electrically before any destructive process is started. Electrical failure analysis involves putting the unit on the ATE and start looking at every single test in the test program to identify any failures in all of the functional blocks. Following that every failure must be characterized over temperature and voltage. This will form the basis of whether the failure is a solid or the failure depends on certain operating conditions. Next step involves creating specific customer patterns that will exercise a failing block and recording failure conditions and status of registers that might give the analyst clues as to where the failure resides.

## 4.5    Summary

Skew lots are essential in identifying product sensitivities and how to address them. The other reason to create skew lots is to characterize your product over all the process variations that will occur during its life cycle. Discussions with circuit designers and process engineers will result in a more appropriate and complete skew table for your product. In general skew lots will take longer to process in the fabrication line so plan accordingly. Analyzing the skew lot yield and comparing it with your yield model is essential in determining where to focus your efforts for yield improvement and product characterization.

Developing a yield model that takes into account the areas of different blocks of the product will enable a closer look at which blocks are contributing to the overall yield loss and how to approach yield improvement activities.

Failure analysis methods and urgency vary depending on the type and stage of failure observation. Appropriate failure analysis techniques must be determined and used to identify root causes quickly.

# Production Silicon Introduction, Volume Ramp and Cost Reduction Phase

5

Let's dive into what are the top items to be considered in this phase of the product life cycle.

## 5.1  Get Ready to Distribute Units to Various Groups

Different organizations in the company would like to get their hands on this revision very soon after it arrives. So start by sending out queries to the different teams and ask them to provide their requirements in the form of an organized table similar to Table 5.1.

Different organizations will of course have vastly different requirements. Reliability groups may need multiple lots of typical silicon for their remaining tests. Lead customer(s) may require corner (skew) units as part of their validation and qualification work.

After you have such a table it will be clear how many wafers need to be started to meet the demand on a timely basis.

PE must understand with pretty good accuracy when the date and time when silicon will arrive at headquarters. This is a critical time for the whole project's success so be as accurate as possible.

Once you have a good enough estimated date, you must make sure that all equipment and personnel is reserved from that date for at least a couple of weeks to meet all the requirements for testing, unit distribution and many other activities. With this information in your pocket you can move on to other activities. Do keep an eye on any changes to the silicon arrival date as it will keep moving around but should converge as it gets closer to endpoint process stages.

Most likely the very first units built will be from an unsorted wafer. This will lead to lower package unit yield so must plan for that to plan for it so you don't fall short of demand and disappoint many teams.

© The Author(s), under exclusive license to Springer Nature Switzerland AG 2022    91
F. Barman, *Semiconductor Product Engineering, Quality and Operations*,
Synthesis Lectures on Engineering, Science, and Technology,
https://doi.org/10.1007/978-3-031-18030-9_5

**Table 5.1**  Sample unit distribution request spreadsheet

| Organization | TT corner | FF corner | SS corner | SF corner |
|---|---|---|---|---|
| Design | 15 | 15 | 15 | 15 |
| Test | 20 | 10 | 10 | 10 |
| Application | 5 | 5 | 5 | 5 |
| SW/FW | 5 | 5 | 5 | 5 |
| Product | 30 | 20 | 20 | 20 |
| Reliability | 100 | 5 | 5 | 5 |
| Lead customer | 50 | 5 | 5 | 0 |
| Total | 225 | 65 | 65 | 60 |

Once silicon arrives, quickly test a few units with a preliminary test program and distribute them to the design and software team so they can start validation activities without any delay. The test coverage will be lacking but urgency to start activities is much more important at this stage. Let teams know what the test coverage is for these first units.

Next activity to focus on is the delivery of revision B units to lead customers. About 2 to 3 weeks after the first lot arrives, a constant flow of units from the subsequent wafer lots must start arriving for test. Before these lots arrive at test a preliminary version of the test program should be in place so there is no delay in testing and shipping to key customers. The preliminary test program should test all major design blocks at all datasheet voltage corners so that a fairly decent level of quality is delivered to the customer.

## 5.2    Have Major Yield Issues Discovered on ES Silicon Been Addressed Adequately?

In parallel with the blind build activities other wafers from the lot will be going through wafer sort in order to meet early production unit demand.

At this point PE must be closely following the wafer test bring up activities to get a first reading of whether yield detractors identified in revision A have been addressed.

Compare the ES and Production wafer test failure parents as soon as it is available. What do you see? Memory failures were 7% in revision A. What is it now?

PLL failures were 3% in revision A. What is that number in revision B?

Many people will want to know the answer to these questions.

Compare wafer yield maps and illustrate major improvements clearly.

Revision A silicon had 12% ATPG failures at minimum voltage? What is that number in revision B?

## 5.3   Have Functional, Performance and Reliability Issues Been Adequately Addressed?

There were several functional issues with ZDD421-A silicon.

Have those issues been addressed? Design team must look at all of these issues and report their findings of fixes, improvements or no change.

Test team very likely disabled one or more tests in ZDD421A test programs to bypass these functional issues.

If fixes in design worked then the disabled tests should be enabled and characterized over PVT.

ZDD421A was not able to meet 1 GHz top speed specification. How does revision B compare in performance?

Create data visualizations that clearly demonstrate any appreciable yield and performance improvements.

## 5.4   Are There Any Reliability Test(s) that Need to Be Re-Done Due to Failures in ES Silicon?

ZDD421A high speed IO bank failed to meet 1.5 K human body model (HBM) ESD stress tests. How does revision B silicon compare in that respect? PE must gather the data and make a report showing improvements have worked and those IOs can now meet 1.5 kV ESD tests. Usually if HBM ESD stress does not meet 1.5 kV requirements the CDM will also follow suit. Therefore the CDM stress testing is also required to make sure all the improvements incorporated into revision B (production silicon) have improved the CDM ESD stress performance.

During reliability test on revision A silicon there were valid failures observed due to HTOL stress test after 500 h. That means all or a portion of the HTOL test must be repeated to ensure the failure modes have been addressed. In general if there are failures at any of the read points after HTOL stress there must be detailed fault isolation and fixes implemented on revision B. This is one of the most challenging failure analysis activities and takes the longest to produce results. In general HTOL must be re-done on at least one lot of revision B silicon with no failures before it can be concluded that the fixes have the intended results.

## 5.5   Test Program Content and Flow Must Be Finalized

Test programs for production testing of wafer and package level tests are much more comprehensive compared to ES phase programs. These test programs will do the work of screening out parts that don't meet one or more datasheet specifications.

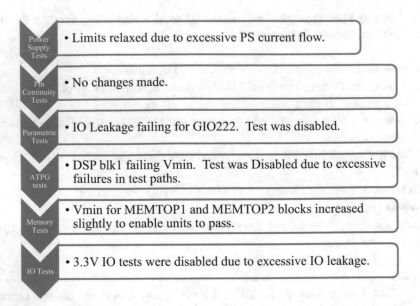

**Fig. 5.1**  Typical test program for wafer test in ES phase

The test programs for wafer, package and QA must be progressing towards more and more tests and more complete fault coverage. They will include additional functional patterns, more complex memory test algorithms and tighter parametric limits and voltage guard bands.

A typical test program for wafer test in ES phase looks like Fig. 5.1.

Due to issues in the chip design or possibly the testing methods these adjustments had to be made to enable the test program to produce passing units that although do not meet datasheet specifications are good enough to ship to lead customers as "engineering samples". These issues must be communicated to the design team so they can be addressed in production revision of silicon. In general these adjustments will also need to be communicated to the lead customer so they are aware of the shortcomings and can make adjustments to their system to accommodate these changes.

The production silicon wafer test program will have more tests enabled and limits set correctly as long as all the problems have been either addressed in chip design or testing methodology.

All tests are enabled in revision B test program and tests have minimum and maximum voltage corners enabled (Fig. 5.2).

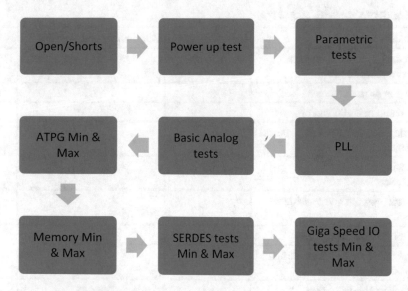

**Fig. 5.2**  Typical high volume test program structure

## 5.6    BOM, Mfg. Flow and Supply Chain Must Be Finalized

Bill of Materials (BOM) is what defines the process steps and materials used to build the final packaged unit. This is a very efficient method to document and communicate these specifications to packaging and test subcontractors.

The material typically include which revision of silicon, packaging materials used during assembly and any mechanical methods used to cut out silicon die from the whole wafer (either mechanical sawing of scribe lines, laser grooving or a combination of both) or packaged units. There are many other processes or material required to assemble silicon wafers into packaged units so the list has to be comprehensive for the particular product being assembled.

Package level stress tests done During ES phase could expose some weaknesses. If there are failures they have to be looked at closely and quickly to determine if any package level processes or material have to change to address related failures.

Typical BOM for a semiconductor based product will look like Table 5.2

This is an important document as it defines and records all materials, processes, testing, packing, labeling and shipping requirements for the finished product.

Since there are many processes and material involved in semiconductor manufacturing it is essential to document all the details as product record and enable any future personnel working on the product to quickly find these essential information.

**Table 5.2** BOM Table for ZDD421A

| Material/Process | Manufacturer | Part number and Revision | Notes |
|---|---|---|---|
| Silicon design part and revision numbers and mask list | | | |
| Fabrication process name | | | |
| Solder bump material and process | | | |
| Wafer thinning process and thickness requirements | | | |
| Lot packing and shipping requirements | | | |
| Wafer test program and revision number | | | |
| Wafer mount and saw process | | | |
| Package Substrate Laminate design and revision number | | | |
| Die attach method, temperature and curing time | | | |
| Underfill material manufacturer and compound number | | | |
| underfill method | | | |
| Curing time and temperature | | | |
| Package Lid | | | |
| Solder ball material composition and attachment method | | | |
| Tray and strapping technique requirements | | | |
| Vacuum seal method and inner box labeling requirements | | | |
| Padding and packing requirements for sealed trays | | | |
| Outer box labeling requirements | | | |
| Packaged unit manufacturing flow document | | | |
| Final Test flow test program name and revision numbers | | | |
| QA test program name and revision | | | |

During the lifetime of any semiconductor based product it is common for materials, processes, test programs and manufacturing subcontractors to be changed for various reasons. The reasons for these changes are varied but usually boil down to cost reduction or improved quality. Some examples of common changes are assembly/test site transfers, wafer fabrication facility change and packaging design or material changes.

Every time a process or material is modified, changed or updated the BOM must be updated to reflect these changes. In general this record is kept in a life cycle management tool such as Agile. Any updates to an approved BOM record must go through review and approval by various organizations before it is fully approved and released. Such review enables clear communication, review and input on any major or minor changes by all the organizations involved.

Process changes are generally categorized as minor or major. Change in one metal layer to reconfigure a path will likely be categorized as a minor change. Whereas a change in poly and multiple metal layers to modify a large portion of a block will likely be categorized as a major change. The categorization is important because minor changes in a design will not require a PCN (Process Change Notification) to be issued to major customers. Major changes to a chip will require you to issue a PCN and go through a new cycle of reliability qualification and characterization activities. Major changes are costly endeavors and will chew up and delay release of product to volume ramp stage. Changes have to be reviewed and agreed upon by all groups involved in product development to ensure all parties agree on the required changes and are onboard to execute such a lengthy and costly activity.

## 5.7   Are Subcontractors, Suppliers and Supply Chain Ready for Production Silicon Introduction?

While in ES phase ZDD421A was built using the suppliers that are determined to be the best sources to meet product requirements. Reliability stress tests are performed on the packaged units in order to expose any weaknesses in material or processes of those specific suppliers. Depending on the results and findings of the reliability stress tests one or material, processes or in extreme cases, a supplier may have to be changed to fix the weaknesses exposed. Depending on the severity of issues, in most cases, changes in materials or processes will fix issues and it will not be necessary to abandon one subcontractor for another. Bring up of a new supplier at this advanced stage of product development may introduce other variables and cause different set of issues that could ultimately doom the product's chances of success.

Reputable suppliers of semiconductor package assembly will do their own stress tests, specific measurements and track yields and manufacturing issues of ZDD421-B and report any results that could give the team more confidence in releasing the product into the market.

## 5.8    Production Revision Lots Tracking, Yield Analysis and Any Process Targeting

ZDD421B wafer starts plan.

Based on the results of their evaluations of revision A silicon, the team has very high confidence in production revision silicon meeting all datasheet specifications and function as intended. So after many discussions and meetings the team decided to go for the plan outlined in Table 5.3 to meet all the various demands from internal groups as well as key customers.

These lots will have to be tracked very closely down to by wafer in most cases. Besides schedule tracking, the yield numbers, product performance and other data extracted from these lots will reveal where the wafer fabrication process will be targeted to optimize quality, performance and cost structure for the rest of the product's life. These 150 wafers will determine how successful ZDD421 will be for the company and in the marketplace.

Every lot in this plan has a backup to lower the risk of wafer or package unit loss during wafer fabrication and assembly processes. The number of wafers started is usually 25 but in some cases suppliers may agree to smaller lot quantities such as 12 or 13.

The lots labeled as MV lots are the ones that will be put through the proposed manufacturing flow at qualified suppliers and subcontractors. Once the package units are available they must go through the proposed test flow at subcontractors. Test data, wafer maps and

**Table 5.3** Initial volume ramp material planning spreadsheet

| Lot name | # of wafers | Purpose |
|---|---|---|
| Lead silicon | 25 | Initial builds |
| Backup silicon | 25 | Backup for lead lot |
| Skew #1 | 25 | Characterization, Validation etc |
| Skew #2 | 25 | Backup for lead skew lot |
| Risk #1 | 25 | Early customer demand |
| Risk #2 | 25 | Next quarter demand |
| Manufacturing validation (MV) lot#1 | 25 | Pipe cleaner for volume ramp |
| MV lot #2 | 25 | Customer demand |
| MV lot #3 | 25 | Customer demand |
| Risk #3 | 50 | Customer demand |
| Risk #4 | 50 | Customer demand |

yield must be collected and analyzed to ensure that the quality level achieved with the proposed manufacturing flow meets company and customer requirements.

Sufficient quantity of hardware is also required to be at subcontractors to ensure all activities involved are executed on a timely manner. Almost all semiconductor test hardware including testers, hardware for wafer, package test and burn-in boards have procurement lead times of anywhere from 2 to 3 months. Therefore all the hardware needed has to be at subcontractors well ahead of start date of any activities.

All the data (e-test, wafer, final and QA tests) from various manufacturing steps must go directly into the yield management system for future analysis and correlation of key product performance characteristics to wafer fabrication and package assembly process parameters.

## 5.9   Case Study

First two lots of ZDD421B have completed fabrication, assembly and the final test data shows significant yield loss for a logic block critical path Fmax at 105c temperature. Expected yield loss for Fmax test is 1–5% but certain wafers show 2–20% fallout (Fig. 5.3).

How would you approach the issue and formulate a plan with the ultimate goal of finding the root cause and putting in place a fix?

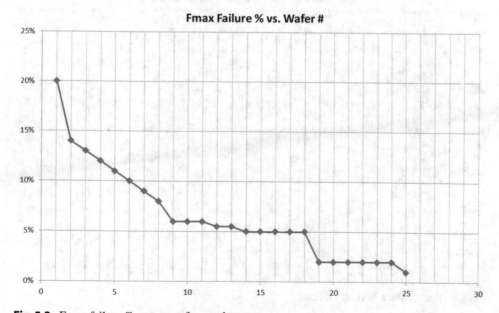

**Fig. 5.3**  Fmax failure % versus wafer number

Here is how I would approach the problem assuming one wafer was packaged as a unique lot in assembly.

Gather a sample of ten units that fail Fmax from one of the lots and get on the ATE with the units to first validate the failures with the test program and temperature forcing unit.

Once the failures are validated, collect Fmax data for these units at 5 temperatures − 45, 0, 25, 90 and 110c to establish the effect of temperature on Fmax.

Collect Fmax data for these units at different core supply voltages (Min, Typical and Max and some values in between).

A strong dependency on voltage may point to an internal voltage regulator issue supplying this particular block or excessive IR drop internal to one of the blocks along the critical path.

A good place to start would be to discuss the above results with the lead design engineer and get his/her opinion on what process parameters are key for high performance of this critical path.

Let say the discussion resulted in a conclusion that absolute value of P-channel transistor Idsat is a critical factor for this particular path. Use your YES capabilities to chart Fmax versus |PIdsat|. A hypothetical chart would look like Fig. 5.4

This chart clearly shows that the process must be centered to stay away from the cliff at ~420 uA/um.

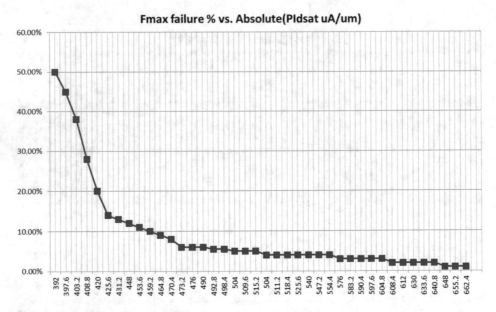

**Fig. 5.4**  Fmax failure % versus |PIdsat|

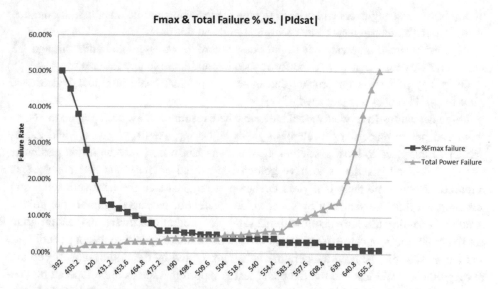

**Fig. 5.5** |PIdsat| versus Fmax failure rate and total power failure rate

This is a simple illustration but in reality there maybe multiple factors in play. Maybe NIdsat may have some effect on Fmax of this critical path!

Other useful charts are how Fmax is affected by voltage and temperature? These results may point to a problem with temperature control during testing of each unit. If the junction temperature of units is not well controlled during testing, it is possible that it may rise gradually and result in degradation in performance or increase in power consumption.

From Fig. 5.5 it can be concluded that the fabrication process must be centered around a |PIdsat| of 550–600 uA/um to minimize Fmax failure rate. However if the PE superimposes the failure rate of total chip power in the same plot a more nuanced analysis and result emerges. Please take a look at Fig. 5.5 and note the increasing power failure rate as |PIdsat| is increased.

Therefore once the effect of increasing P-channel transistor current drive on power consumption is presented the desirable target range of |PIdsat| is tightened to 490–550 uA/um. A seasoned PE would know that such trade offs often exist and must be looked at carefully.

## 5.10   Assess the Quality and Reliability Achieved Using the Proposed Production Manufacturing Flow

First assessment results come when the MV lot package tests have completed and the lots are going through QA tests. If any failures are observed at QA stage they must be analyzed, root cause identified and corrective actions put in place to prevent reoccurrence.

In most cases test engineers will need to tighten the test voltages and other conditions such as timing or parametric tests so that marginal units do not pass.

Another essential assessment is to put one or more lot of assembled units through the same reliability tests that were done for revision A and observe any failures.

Other key indicators for product and quality of product health are actual yields vs. what the yield model has predicted.

Large deviations from yield model indicate where resources must be deployed to understand and put in place fixes to increase yields and bring them up to what yield model predicts. First place to look is whether there are any hardware issues with that particular test step. A good practice is to have golden wafers and golden units that are used as reference. Every time there is a yield deviation at any test step the hardware setup and test program must be validated by using golden wafers or packaged units. If the golden material is passing the same setup with the same test program hardware and test program are likely not the root cause. Next would be to look at the temperature forcing conditions and ensure that the intended temperature is being reported by the failing unit. If you are forcing 90c at wafer test and the die under test is reporting 110c as the temperature, you have to look at why and fix it. Also if the test program is supposed to apply 0.7 v for a certain supply but the internal sensing is reporting 0.6 v you have to look at where there is excessive IR drop in the testing environment.

## 5.11  Final Manufacturing Flow Must Be Determined During This Phase

What will be the manufacturing process for ZDD421-B during the volume ramp phase?

At the early stages of volume ramp ZDD421-B has a fairly complex manufacturing flow including multiple wafer, package and reliability tests. The reason for this complex flow is that during the ES phase it was determined that all these tests capture unique failures that must be sorted out.

The product engineer's main job at this point of development is to capture and analyze these supposedly unique failures and find a way to identify them at earlier process stages using a combination of voltages and other test conditions that are already present in the flow.

For example, a good question to ask is, can the test identify 0c wafer test failures at 25c wafer test or 25c package level test?

These types of analysis will involve packaging die or dice that failed at wafer test level and analyze them in detail to understand the nature of the failure and other ways to expose this failure.

A manufacturing flow that includes burn-in is an expensive proposition. Burn-in in high volume manufacturing will require packaged units to be diverted into burn-in stress very similar to what was done for HTOL but with a much shorter duration (for example 24 or

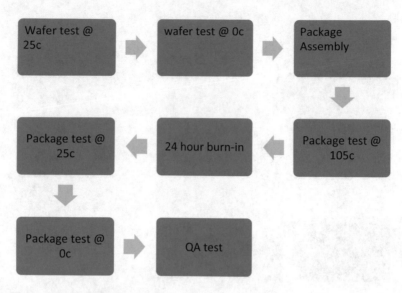

**Fig. 5.6**  Not optimized manufacturing flow

48 h). After burn-in is completed there will be a test insertion to assess whether the stress has resulted in performance degradation and/or functional failure in any units. The most stringent requirement would be to burn-in 100% of packaged units! As you can imagine this would require a large number of burn-in boards, additional time to process 100% of units through this step. A less stringent requirement might be to burn-in a sample of the units from each manufacturing lot. The higher the failure rate during HTOL portion of the reliability qualification stress, the higher the percentage of parts will need to go through production burn-in. Therefore it is of utmost importance to have clean results from the HTOL during reliability qualification activities. In order to remove a wafer test insertion the PE must demonstrate that there will be minimal effect on down the stream packaged unit tests such as final or QA. PE must do a thorough cost/benefit analysis and show that the savings from wafer test elimination are much higher than the cost of additional failures at package level tests before management approves elimination of a wafer test insertion (Figs. 5.6 and 5.7).

## 5.12  Publish Final Characterization Report

Dust off the characterization report you put together in the ES phase and collect the data needed to supplement this for final publication. Start working with a test engineer so that he/she can plan for B revision of product characterization test program. The output data format must be in STDF format so that it can be easily imported into your data analysis

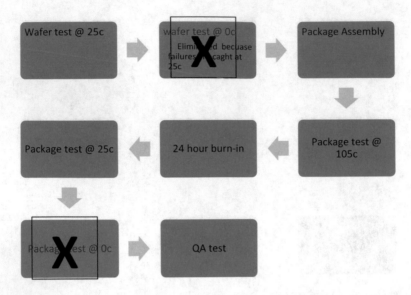

**Fig. 5.7** Optimized manufacturing flow

program. Please refer to the characterization Chap. 10 for further details. Start the final report with an executive summary to frame the results and conclusions upfront.

Use charts to visualize and establish relationships between and observation and a possible major factor. You can use histograms to visualize large volume of data for a certain performance characteristic of the chip.

Upfront in your report present and discuss any parameters that don't meet the Cpk guidelines (presented in Chap. 3) over PVT. Present possible solutions to increase Cpk to an acceptable level for each of those parameters.

Final characterization report must also include an updated critical parameter shift analysis. This data is collected during each read point of the HTOL stress test. For example if Fmax shows a degradation of 100 MHz after 1000 h of stress, at least a 100 MHz guardband must be added to the test program limit to account for this shift. As an example, if Fmax for a particular path has a minimum limit of 500 MHz, it must now have a minimum of 600 MHz Fmax at time zero. After 10 years of normal operation in the filed the performance may drop from 600 to 500 MHz but still meets the minimum datasheet specification.

## 5.13  Publish Final Reliability Qualification Report

In this report you will publish the final results of the qualification reliability tests performed on revision A and revision B. In general, tests that fully passed on revision A

**Table 5.4**  Summary reliability report format

| | Failure count/Total count | | | | | | | | |
|---|---|---|---|---|---|---|---|---|---|
| | HTOL | BIASED HAST | ESD | LU | HAST | TMCL | BAKE | DROP | L2 Qualification |
| Lot #1 | 0/77 | 0/25 | 0/10 | 0/10 | 0/25 | 0/25 | 0/25 | 0 failures | 0/50 |
| Lot #2 | 0/77 | 0/25 | 0/10 | 0/10 | 0/25 | 0/25 | 0/25 | 0 failures | 0/10 |
| Lot #3 | 0/77 | 0/25 | 0/10 | 0/10 | 0/25 | 0/25 | 0/25 | 0 failures | 0/10 |

silicon do not have to be redone on revision B. However, some of your more demanding customers may require redo of all reliability tests for revision B. It is possible to negotiate such requirements but you will need to present a very convincing argument to win agreement for dropping any requests. Start the final report with an executive summary to frame the results and conclusions upfront.

Then move on to reporting the top line results in a table format similar to Table 5.4

If there are any valid failures for any of the stress tests there must be a report provided for root cause, containment actions, corrective actions and validation of corrective actions before a customer will be happy with the reliability qualification report.

## 5.14   Test Cost Reduction

There are three main factors in IC test cost. Product test time, how many units can be tested at the same time and Tester Cost per hour.

## 5.15   Test Time Reduction

Eliminate tests that have zero failures per one million units.

Reduce test time of functional tests by increasing clock speed. There may be issues here since not all circuits are designed to perform at higher clock speeds.

Reduce any unnecessary wait times at beginning or end of a functional test.

Test Pin Parametric (such as leakage, and connectivity) in parallel rather than one pin at a time.

## 5.16    Test Step Elimination

Study failures at one step with an eye to detect and eliminate them in the previous test step. This process basically pushes failures from one test step to an earlier test insertion. QA failures are pushed back to final test and final test failures are pushed back to wafer test insertions.

Increase wafer test coverage and stresses to eliminate die that will fail at package test. You will save both packaging cost and test time incrementally. It is common for products manufactured with advanced CMOS technologies to require a short voltage stress test at wafer test to eliminate any weak units that may fail at package level tests. It is critical to set a safe level of voltage stress for these types of tests so that healthy units are not damaged due to the stress. You only want to eliminate weak units. This may require several iterations to fine tune the stress level.

## 5.17    Manufacturing Step Elimination

Revision A required 100% burn-in because the product showed excessive failures during HTOL. These failures were analyzed and corrected in revision B. Verify this by redoing HTOL. If there are no failures during HTOL of revision B you can eliminate 100% burn-in and replace it with a sample burn-in. Once enough samples are run with no failures PE can eliminate sample burn-in as well.

## 5.18    System Level Test Elimination

SLT is extremely expensive due to very long test times and manual test insertions. It also contributes to degradation of quality due to manual handling of individual units. Eliminating this test will result in huge reduction in product cost and overall cycle time for manufacturing.

First step is to increase the yield at SLT to above 99%. Once you do that give yourself a pat in the back! If you eliminate SLT at this point you will add an additional 10,000 DPPM to the overall product failure rate. This is not acceptable to the company and customers. Once SLT failure rate is close to 1,000 DPPM (99.99% or higher yield) PE can start looking at individual failures and how to screen them out at wafer or package tests. Once new tests to screen these small numbers of failures at package or wafer test are added more lots must be run to validate that no LT failures are detected. A good sample size would be 20,000 units.

Table 5.5 shows the impact of several cost reduction activities that netted nearly 50% savings in product cost.

**Table 5.5** Impact of cost reduction actions on final product cost

|  | Original cost structure | Cost reduced structure | Cost reduced structure |  |
|---|---|---|---|---|
| Overall yield | 61% | 70% | 70% |  |
| NDPW | 350 | 403 | 403 |  |
| Wafer cost | $7,000.00 | $7,000.00 | $7,000.00 |  |
| Wafer test cost | $100.00 | $100.00 | $100.00 |  |
| Tested die cost | $20.30 | $17.60 | $17.62 |  |
| Packaging cost | $5.00 | $5.00 | $2.50 |  |
| Package unit cost | $25.30 | $22.60 | $20.12 |  |
| Package test1 cost | $2.00 | $1.50 | $1.50 |  |
| Package test2 cost | $1.00 | $0.00 | $0.00 |  |
| BURN-IN COST | $3.00 | $0.00 | $0.00 |  |
| Test yield | 90% | 99% | 99% |  |
| System level test cost | $5.00 | 0% | 0% |  |
| System level test yield | 90% | 100% | 100% |  |
| Fully tested unit cost | $44.20 | $24.34 | $21.84 |  |
| Finish operations cost | $1.00 | $1.00 | $1.00 | Savings |
| Ready to ship product cost | $45.20 | $25.34 | $22.84 | 49% |

## 5.19   2nd Source Fab and Assembly Bring Up Is Close to Bring Up and Volume Ramp

Once validated volume projections exceed plan by a significant number it is time to explore 2nd source for wafer fabrication as well as assembly and test sites.

Wafer fabrication 2nd source poses its own set of challenges. The concept of "process matching" becomes critical element here. Process matching is a methodology by which a predetermined set of product characteristics are compared between primary source and 2nd source fabrication facilities. These characteristics must match between primary and second source. There is a small tolerance allowed after which the specific parameter is declared "not matching". Yield and quality matching studies must be performed with product manufactured using 2nd source wafers to ensure that quality and cost are not sacrificed in any way as compared to primary source.

Product engineer will have to basically plan and execute a complete product qualification using wafers manufactured at 2nd source facilities.

The same exercise will have to be done if a 2nd source assembly site is brought onboard. The extent of this qualification could be limited to package qualification related tests outlined in Chap. 10.

In certain cases the qualification of a 2nd source wafer fab and package assembly site could be combined into one plan. The disadvantage with such plan is that it will be more difficult to isolate sources of any failures to wafer or package change.

Introducing second source wafer fabrication and/or package assembly suppliers is a "major change" that will require product change notifications (PCNs) to be issued, reviewed and approved by key customers well ahead of planning and execution. 1 year ahead is a minimum time that customers should be informed f such plans. Most customers would want to have inputs into the qualification planning and execution phase as this greatly affects their manufacturing processes.

## 5.20    Aggressive Yield Improvements Are Your Key to Cost Reduction

PE must look at every test and manufacturing step and squeeze out as much yield improvements as possible. Key to success in this effort is to take a deep dive into dice and units that have failed at wafer or package test steps. Failing dice must be packaged s that it can be looked at from every angle to ensure that failure mechanisms are understood and fixes are put in place. Packaged unit failures should first be looked at from the viewpoint of preventing passing wafer test and then failing at package level where the cost of failure is more expensive because of the package wrapped around it, tested and then thrown into the fail bin. This is contrary to "cost reduction".

Testing at temperatures below 25c are much more expensive and prone to issues due to expensive equipment required to bring temperatures down and keep them stable at those levels. Any moisture in the air will freeze at low temperatures and settle into equipment, unit package or test hardware causing temporary failures or lasting damage.

Avoid low temperature testing as much as possible to prevent such issues from creeping into product's manufacturing flow.

In advanced technologies there exists a phenomenon called the "temperature inversion". This phenomenon describes why in advanced technology nodes performance of products actually degrades (i.e. product actually slows down at cold temperatures). This is the inverse of what happens in older technologies. Products build using older technologies actually increase their performance at cold temperatures).

Ideally the product engineer puts together a test flow that has one insertion at wafer level and one insertion at package level. PEs main task during volume ramp is to collect and analyze massive amounts of data generated to eliminate any additional test insertions.

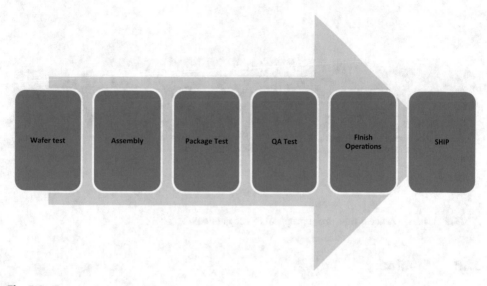

**Fig. 5.8**  Supply chain simplification

Goal:

Reach end of volume ramp stage with the following manufacturing flow validated and implemented.

Further simplify manufacturing flow to beat standard cost models.

Explore alternate material that meets performance, quality and reduces cost.

Supply chain simplification (Fig. 5.8).

Automotive grade products need careful considerations in cost reduction phase.

Every customer in this market segment requires following a change approval process that is presented in Fig. 5.9.

This process could take over one year to complete and would cost the company millions of dollars. At every review or approval stage your automotive customer may decide that it is not worth their time, money or effort to complete this process.

There has to be a very good reason for the customer to approve the proposed change.

Reduction in existing test insertions must be validated by showing that there is no impact to quality. This means that any failures from eliminated test must be caught by an existing test step. Removal of test steps are sometimes looked at positively by customers since they recue the chances of degradation or damage and reduce the total cycle time for product delivery.

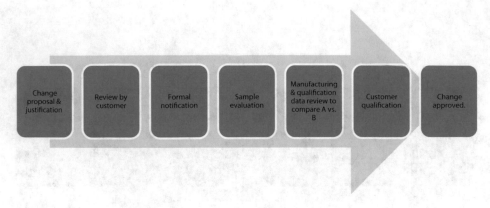

**Fig. 5.9** Change process flow from proposal to acceptance

## 5.21   Summary

During production silicon phase the final revision of the ZDD421A design will be released to the market place once all the modifications and enhancements have been validated by customers and internal teams.

Any reliability stress tests that had failures during engineering sample phase have to be redone using rev. B silicon and validated to have no failures.

Final characterization and reliability report should be distributed to internal teams and key customers. These two reports will demonstrate the reliability and manufacturability of ZDD421B.

During this phase of product life cycle the manufacturing flow, subcontractors and the product BOM has to be finalized and released to various entities involved in manufacturing of ZDD421B (production version).

Cost reduction phase follows with the goal of reducing total product cost without compromising quality levels customers expect. Additional wafer, packaging and test suppliers are qualified and give a share of the volume.